K12 Mathematics
VOLUME 1

Rose Anita McDonnell

Anne Veronica Burrows

Colleen A. Dougherty

Helen T. Smythe

Catherine D. LeTourneau

M. Winifred Kelly

Mary Grace Fertal

Monica T. Sicilia

with
Dr. Elinor R. Ford

Copyright © 2000 by William H. Sadlier, Inc. Also published as *Progress in Mathematics*, © 2000.

Acknowledgments

Every good faith effort has been made to locate the owners of copyrighted material to arrange permission to reprint selections. In several cases this has proved impossible. The publisher will be pleased to consider necessary adjustments in future printings.

Thanks to the following for permission to reprint the copyrighted materials listed below.

"At the Beach" (text only) by Rebecca Kai Dotlich. Copyright © 2000 by Rebecca Kai Dotlich. Reprinted by permission of Curtis Brown, Ltd.

"At the Table" (text only) by Constance Andrea Keremes. Used by permission of the author, who controls all rights.

"County Fair" by Catherine Danielle LeTourneau. Printed with permission of the author.

"Hats" (text only) by Daphne Lister. Copyright © Daphne Lister. Used by permission of the author.

"Math My Way" (text only) by Carol Diggory Shields from LUNCH MONEY by Carol Diggory Shields. Copyright © 1995 by Carol Diggory Shields. Used by permission of Dutton Children's Books, a division of Penguin Putnam Inc.

Excerpt from MISS SPIDER'S TEA PARTY (text only) by David Kirk. Copyright © 1994 by Callaway Editions, Inc. Reprinted by permission of Scholastic Inc.

"The Playground on Euclid Street" (text only) by Christine Barrett. Printed with permission of the author.

Excerpts from RIDDLE-ICIOUS (text only) by J. Patrick Lewis. Text copyright © 1996 by J. Patrick Lewis. Reprinted by permission of Alfred A. Knopf, Inc.

"Rides" (text only) by Ilo Orleans. Reprinted with permission of Karen. S. Solomon.

"Spending Time and Pennies" (text only) by Helen T. Smythe. Printed with permission of the author.

"Stars" (text only) from WIND SONG, copyright © 1960 by Carl Sandburg and renewed 1988 by Margaret Sandburg, Janet Sandburg, and Helga Sandburg Crile, reprinted by permission of Harcourt Brace & Company.

"Ten Tom-Toms" (text only), Unknown.

Anastasia Suen, Tim Mason, Literature Consultants

All manipulative products generously provided by ETA/Cuisenaire, Vernon Hills, IL.

Photo Credits

Janette Beckman: 336, 337, 338, 344, 354, 359, 371, 455, 457.
FPG/ Tanaka Associates: 284.
Image Bank: 442.
Clay Patrick McBride: 25, 34, 35, 36, 46, 55, 59, 61, 67, 74, 89, 95, 100, 102, 122, 123, 125, 127, 135, 149, 155, 158, 165, 169, 175, 177, 185, 191, 195, 196, 211, 213, 214, 215, 219, 221, 228, 229, 237, 242, 246, 247, 248, 249, 256, 267, 270, 276, 277, 287, 289, 295, 298, 300, 305, 315, 328, 329, 350, 377, 381, 385, 387, 388, 393, 398, 399, 415, 416, 418, 419, 420, 423, 424, 425, 435, 441, 442, 445, 447, 448, 455, 467, 465.
Ken Karp: 66, 94, 124, 193, 251, 292, 427, 461.
Stock Market/ Craig Tuttle: 400 top center;
Ariel Skelley: 400 center left.
Tony Stone Images: 400 top left, 400 bottom left; *Mike McQueen:* 397 bottom left; *Greg Probst:* 400 top left; *Roy Gumpel:* 400 center right; *Stewart Cohen:* 400 bottom right.

Illustrators

Bernard Adnet
Don Bishop
Lisa Blackshear
John Corkery
Deborah Drummond
Rob Dunlavey

Laura Freeman
Tim Haggerty
Steve Henry
Gary Johnson
Dave Jonason
David Julian

Andy Levine
Jason Levinson
Ellen Joy Sasaki
Stephen Schudlich
Matt Straub
Blake Thornton

George Ulrich
Richard Weiss
Nadine Bernard Westcott
Dirk Wunderlich

Copyright © 2000 by William H. Sadlier, Inc. Also published as *Progress In Mathematics,* © 2000. By arrangement with the publisher.

This publication, or any part thereof, may not be reproduced in any form, or by any means, including electronic, photographic, or mechanical, or by any sound recording system, or by any device for storage and retrieval of information, without the written permission of the publisher. Address inquiries to Permissions Department, William H. Sadlier, Inc., 9 Pine Street, New York, NY 10005-1002.

S is a registered trademark of William H. Sadlier, Inc.

Edition specially produced for K12, Inc.—http://www.k12.com for more information.

Printed in the United States of America

ISBN: 0-8215-2601-Z
123456789/05 04 03 02 01

Contents

Letter to Family ix
Skills Update Magazine (For review of
essential and previously learned skills) x
Introduction to Problem Solving 13

CHAPTER 1
Numbers to 12

Theme: Bugs 15
Math Alive at Home 16
  Miss Spider's Tea Party by David Kirk
Math Connections: Real World 17
Cross-Curricular Connections:
  Language Arts 18
 1-1 Write One and Two 19
 1-2 Write Three and Four 21
 1-3 Write Five and Zero 23
*1-4 One More, One Fewer 25
 1-5 Pictograph 27
 1-6 Write Six and Seven 29
 1-7 Write Eight and Nine 31
 1-8 Write Ten 33
 1-9 Write Eleven and Twelve 35
 1-10 Zero to Twelve 37
 1-11 Order 0–12 39
 1-12 Counting on a Number Line 41
 1-13 Counting Back 43
 1-14 Before, After, Between 44
*1-15 Compare 45
 1-16 Odd and Even 46
 1-17 Ordinals 47
*1-18 *Problem-Solving Strategy:*
      Find a Pattern 49
*1-19 *Problem-Solving Applications:*
      Draw a Picture 51
Chapter Review and Practice 53
Math-e-Magic: Fewest and Most 54
Performance Assessment 55
Check Your Mastery 56

CHAPTER 2
Understanding Addition: Facts to 6

Theme: Pet Store 57
Math Alive at Home 58
  "Hats" by Daphne Lister
Math Connections: Modeling Figures ... 59
Cross-Curricular Connections: Science ... 60
 2-1 Understanding Addition 61
*2-2 Addition Sentences 63
 2-3 Sums of 4 65
*2-4 Sums of 5 67
*2-5 Sums of 6 69
*2-6 Change the Order 71
      Do You Remember? 73
*2-7 Addition Patterns 74
 2-8 Technology:
      Learn about Computers 75
 2-9 *Problem-Solving Strategy:*
      Use a Graph 77
*2-10 *Problem-Solving Applications:*
      Act It Out 79
Chapter Review and Practice 81
Math-e-Magic: Addition with Money 82
Performance Assessment 83
Check Your Mastery 84

* **Algebraic Reasoning**

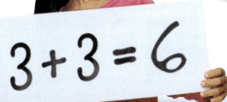

CHAPTER 3
Understanding Subtraction: Facts to 6

Theme: School Days		85
Math Alive at Home		86
"At the Table" by Constance Andrea Keremes		
Math Connections: Visual Reasoning		87
Cross-Curricular Connections: Social Studies		88
3-1	Understanding Subtraction	89
*3-2	Subtraction Sentences	91
3-3	Subtract from 4	93
3-4	Subtract from 5	95
3-5	Subtract from 6	97
*	Do You Remember?	99
*3-6	Subtraction Patterns	100
*3-7	Relate Addition and Subtraction	101
*3-8	Zero in Facts	102
*3-9	*Problem-Solving Strategy: Choose the Operation*	103
*3-10	*Problem-Solving Applications: Write a Number Sentence*	105
	Chapter Review and Practice	107
*	**Math-e-Magic: Missing Operations**	108
	Performance Assessment	109
	Check Your Mastery	110
	Cumulative Review I	111

CHAPTER 4
Addition Facts to 12

Theme: Transportation		113
Math Alive at Home		114
"Rides" by Ilo Orleans		
Math Connections: Sort and Compare		115
Cross-Curricular Connections: Language Arts		116
*4-1	Sums of 7	117
*4-2	Sums of 8	119
*4-3	Sums of 9	121
4-4	Sums of 10	123
*4-5	Sums of 11	125
4-6	Sums of 12	127
*4-7	Number-Line Addition	129
*4-8	Add: Use Patterns	131
*4-9	Doubles and Near Doubles	133
*4-10	Add Three Numbers	135
*4-11	Chain Addition	137
4-12	Addition Strategies	139
	Do You Remember?	140
4-13	Input-Output	141
*4-14	*Problem-Solving Strategy: Ask a Question*	143
4-15	*Problem-Solving Applications: Choose the Operation*	145
	Chapter Review and Practice	147
*	**Math-e-Magic: Missing Addends**	148
	Performance Assessment	149
	Check Your Mastery	150

* **Algebraic Reasoning**

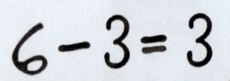

CHAPTER 5
Subtraction Facts to 12

Theme: County Fair 151
Math Alive at Home 152
    "County Fair"
    by Catherine Danielle LeTourneau
Math Connections: Tally and Graph 153
Cross-Curricular Connections:
    Map Reading 154
* 5-1   Other Names: Sums and Differences 155
* 5-2   Subtract from 7 157
  5-3   Subtract from 8 159
  5-4   Subtract from 9 161
  5-5   Subtract from 10 163
* 5-6   Subtract from 11 165
  5-7   Subtract from 12 167
* 5-8   Number-Line Subtraction 169
* 5-9   Subtract: Use Patterns 171
  5-10 Subtraction Strategies 173
    Do You Remember? 174
* 5-11 Check by Adding 175
* 5-12 Fact Families 177
* 5-13 *Problem-Solving Strategy:*
    Draw to Compare 179
* 5-14 *Problem-Solving Applications:*
    Use a Model 181
Chapter Review and Practice 183
* **Math-e-Magic: Odd and**
    Even Sums and Differences 184
Performance Assessment 185
Check Your Mastery 186
Cumulative Test I 187

***Algebraic Reasoning**

CHAPTER 6
Place Value to 100

Theme: Toys and Games 189
Math Alive at Home 190
    "Lunch Money" by Carol Diggory Shields
Math Connections: Making Ten 191
Cross-Curricular Connections:
    Multicultural Studies 192
  6-1   Numbers 10–19 193
  6-2   Ten to Nineteen 195
* 6-3   Count by Tens 197
  6-4   Numbers 20–39 199
  6-5   Numbers 40–59 201
* 6-6   Numbers through 59 203
  6-7   Numbers 60–79 205
  6-8   Numbers 80–99 207
  6-9   Numbers to 100 209
* 6-10 Hundred-Chart Patterns 211
* 6-11 Count On: Number After 213
* 6-12 Count Back: Number Before 214
  6-13 Before, Between, After 215
    Do You Remember? 216
* 6-14 Greater Than, Less Than 217
* 6-15 Count by Fives 219
  6-16 Count by Twos 221
  6-17 Equal Groups 223
  6-18 Sharing 225
  6-19 Separating 227
  6-20 Technology:
    Calculator Skip Counting 229
* 6-21 *Problem-Solving Strategy:*
    Guess and Test 231
  6-22 *Problem-Solving Applications:*
    Logical Reasoning 233
Chapter Review and Practice 235
Math-e-Magic: Counters Left Over 236
Performance Assessment 237
Check Your Mastery 238

CHAPTER 7
Money and Time

Theme: Hobbies and Collectibles 239

Math Alive at Home 240
    "Spending Time and Pennies"
    by Helen T. Smythe

Math Connections: Graphs/Visual Reasoning 241

Cross-Curricular Connections: Language Arts 242

7-1	Nickels	243
7-2	Dimes	244
7-3	Count On by Pennies	245
7-4	Quarters	247
7-5	Count On by Nickels	249
7-6	Count Mixed Coins	251
7-7	Trading	253
7-8	Using Money	255
7-9	Equal Amounts	257
	Do You Remember?	258
7-10	Hour	259
7-11	Half Hour	261
✶ 7-12	Time Patterns	263
7-13	Name the Time	265
7-14	Estimate Time	267
7-15	Calendar	269
7-16	*Problem-Solving Strategy:* Logical Reasoning	271
7-17	*Problem-Solving Applications:* Hidden Information	273

Chapter Review and Practice 275
Math-e-Magic: Trading for 1 Dollar 276
Performance Assessment 277
Check Your Mastery 278
Cumulative Review II 279

✶ **Algebraic Reasoning**

CHAPTER 8
Geometry, Fractions, and Probability

Theme: Playground 281

Math Alive at Home 282
    "The Playground on Euclid Street"
    by Christine Barrett

Math Connections: Position—Inside, Outside, On 283

Cross-Curricular Connections: Multicultural Art 284

8-1	Open and Closed Figures	285
8-2	Sides and Corners	287
8-3	Plane Figures	289
	Do You Remember?	291
8-4	Same Shape	292
✶ 8-5	Same Shape and Size	293
8-6	Symmetry	295
8-7	Slides and Turns	297
8-8	Space Figures	298
8-9	Cube, Rectangular Prism, Pyramid	299
8-10	Cylinder, Cone, Sphere	301
8-11	Make a Bar Graph	303
8-12	Equal Parts	305
8-13	One Half	307
✶ 8-14	One Third	309
8-15	One Fourth	311
8-16	Part of a Set	313
8-17	Always, Sometimes, Never	315
8-18	Arrangements	317
8-19	More or Less Likely	319
8-20	Technology: LOGO	321
8-21	*Problem-Solving Strategy:* Logical Reasoning	323
8-22	*Problem-Solving Applications:* Use Drawings and Models	325

Chapter Review and Practice 327
Math-e-Magic: Figures Inside Plane Figures . 328
Performance Assessment 329
Check Your Mastery 330

CHAPTER 9
Add and Subtract Two-Digit Numbers

Theme: Music/Instruments **331**
Math Alive at Home **332**
    "Ten Tom-Toms" **by Unknown**
Math Connections:
    Number-Line Estimation **333**
Cross-Curricular Connections:
    Multicultural Music **334**
  9-1 Estimate: About How Many 335
  9-2 Add Tens 337
  9-3 Subtract Tens 338
✱ 9-4 Add and Subtract Tens 339
✱ 9-5 Add Tens and Ones 341
✱ 9-6 More Adding Tens and Ones 343
✱ 9-7 Add Tens or Ones 345
  9-8 Subtract Tens and Ones 347
✱ 9-9 More Subtracting Tens and Ones .. 349
  9-10 Subtract Tens or Ones 351
        Do You Remember? 353
  9-11 Estimating Sums and Differences .. 354
✱ 9-12 Add Money 355
  9-13 Subtract Money 357
✱ 9-14 Using Addition and Subtraction ... 359
  9-15 Regroup in Addition 361
  9-16 Regroup in Subtraction 363
  9-17 *Problem-Solving Strategy:*
        Logical Reasoning 365
✱ 9-18 *Problem-Solving Applications:*
        Choose the Operation 367
Chapter Review and Practice **369**
✱ **Math-e-Magic: Letter-Number Code** **370**
Performance Assessment **371**
Check Your Mastery **372**
Cumulative Review III **373**

CHAPTER 10
Measurement

Theme: Vegetable Garden **375**
Math Alive at Home **376**
    Excerpts from *Riddle-icious* **by J. Patrick Lewis**
Math Connections: Mapping Area **377**
Cross-Curricular Connections:
    Language Arts **378**
  10-1 Estimate and Measure 379
  10-2 Perimeter 381
  10-3 Inches 383
  10-4 Feet 385
  10-5 Cups and Pints 387
  10-6 Quarts 388
✱ 10-7 Pounds 389
  10-8 Centimeters 391
  10-9 Using Centimeters 393
✱   Do You Remember? 394
  10-10 Estimate a Liter 395
  10-11 Kilogram 397
  10-12 Measuring Tools 399
  10-13 Temperature 400
  10-14 Technology: Using a Calculator 401
  10-15 *Problem-Solving Strategy:*
         Use a Map 403
  10-16 *Problem-Solving Applications:*
         Logical Reasoning 405
Chapter Review and Practice **407**
Math-e-Magic: Visual Thinking **408**
Performance Assessment **409**
Check Your Mastery **410**

✱ **Algebraic Reasoning**

CHAPTER 11
Addition and Subtraction Facts to 18

Theme: Beach.................................... 411
Math Alive at Home............................ 412
   "At the Beach" by Rebecca Kai Dotlich
✱ **Math Connections: Missing Numbers**.... 413
✱ **Cross-Curricular Connections:**
   Art and Space Figures................. 414
11-1   Sums of 13 and 14 415
11-2   Subtract from 13 and 14............. 417
11-3   Sums of 15 and 16 419
11-4   Subtract from 15 and 16............. 421
11-5   Facts of 17 and 18 423
✱ 11-6   Fact Families 425
         Do You Remember?.................... 426
✱ 11-7   Three Addends............................ 427
11-8   *Problem-Solving Strategy:*
         Extra Information..................... 429
11-9   *Problem-Solving Applications:*
         Draw a Picture........................... 431
       Chapter Review and Practice 433
✱ **Math-e-Magic: Addition Table**............ 434
       Performance Assessment 435
       Check Your Mastery........................... 436
       Cumulative Test II............................... 437

CHAPTER 12
Moving On in Math

Theme: Space and Stars...................... 439
Math Alive at Home........................... 440
   "Stars" by Carl Sandburg
Math Connections:
   Fractions and Probability............ 441
Cross-Curricular Connections: Science... 442
✱ 12-1   Find the Rule 443
✱ 12-2   Number-Sentence Balance........... 444
✱ 12-3   Missing Operations 445
✱ 12-4   Missing Numbers 446
12-5   Regrouping Money 447
12-6   Add Ones: Regroup..................... 449
12-7   Subtract Ones: Regroup.............. 451
✱ 12-8   Three Addends: No Regrouping ... 453
12-9   Hundreds, Tens, Ones................. 455
✱ 12-10 Add and Subtract
         Three-Digit Numbers.................. 457
12-11 *Problem-Solving Strategy:*
         Make a Table............................. 459
✱ 12-12 *Problem-Solving Applications:*
         Guess and Test........................... 461
       Chapter Review and Practice 463
✱ **Math-e-Magic:**
   Graphing Ordered Pairs.............. 464
       Performance Assessment 465
       Check Your Mastery........................... 466

End-of-Book Materials

Still More Practice............................. 467
Mental Math...................................... 478
Picture Glossary Magazine 483

✱ **Algebraic Reasoning**

Dear Family:

We are pleased that your child will be learning mathematics with us.

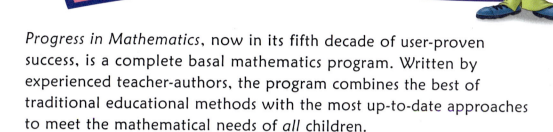

Progress in Mathematics, now in its fifth decade of user-proven success, is a complete basal mathematics program. Written by experienced teacher-authors, the program combines the best of traditional educational methods with the most up-to-date approaches to meet the mathematical needs of *all* children.

The scope and sequence of *Progress in Mathematics* meets the new Standards. Using this program, students may progress as quickly as they can or as slowly as they must.

First-grade topics to be studied include: number recognition, addition, subtraction, place value, time, money, geometry, measurement, fractions, graphing, and introductory concepts in statistics and probability. Special attention is given to critical thinking, problem solving, and the use of technology.

But overall success in achieving the goals of this program is dependent upon active teacher-family-student interaction. You can help your child achieve her/his maximum learning level in mathematics by: (a) talking to your child about mathematics in everyday situations, (b) encouraging your child so that she/he will like mathematics, (c) providing quiet space and time for homework, and (d) instilling the idea that by practicing concepts she/he can have fun while learning mathematics.

Throughout the year, your child will bring home *Math Alive At Home* pages for you and your child to complete together. These pages include fun-filled activities that will help you relate the mathematics your child is learning to everyday life.

We know that by using *Progress in Mathematics* your child will learn to value math, become confident in her/his ability to do math and solve problems, and learn to reason and communicate mathematically.

Sincerely,

The Sadlier-Oxford Family

A Review of Mathematical Skills from Kindergarten

CONTENTS

Same or More	1
Fewer	2
1 to 4	3
5 to 7	4
8 to 10	5
Count by 1: Pennies	6
Join	7
Separate	8
Positions	9
Calendar	10
Same, Longer, Longest	11
Same, Shorter, Shortest	12

This magazine belongs to _____.

I can match. Then I can ring the group with **more**.

Draw a group with more.

Same or More

one 1

_____ can match.

I can ring the group with **fewer**.

Draw a group with fewer.

2 two   **Fewer**

_____ can count to 4.

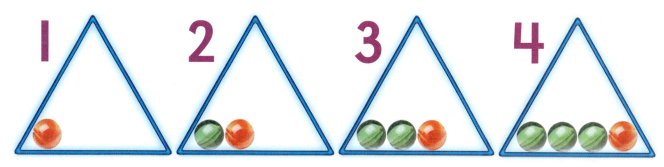

Ring how many.

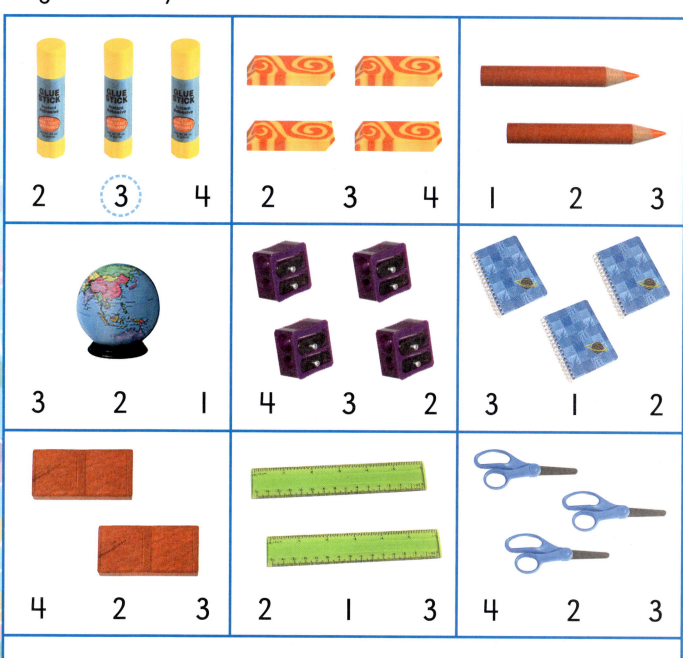

Draw 3 ☺.

1 to 4   three 3

_____

- -

_____ can count to 7.

Ring how many.

Draw more than 4 .

5 to 7

_____ can count to 10.

Ring how many.

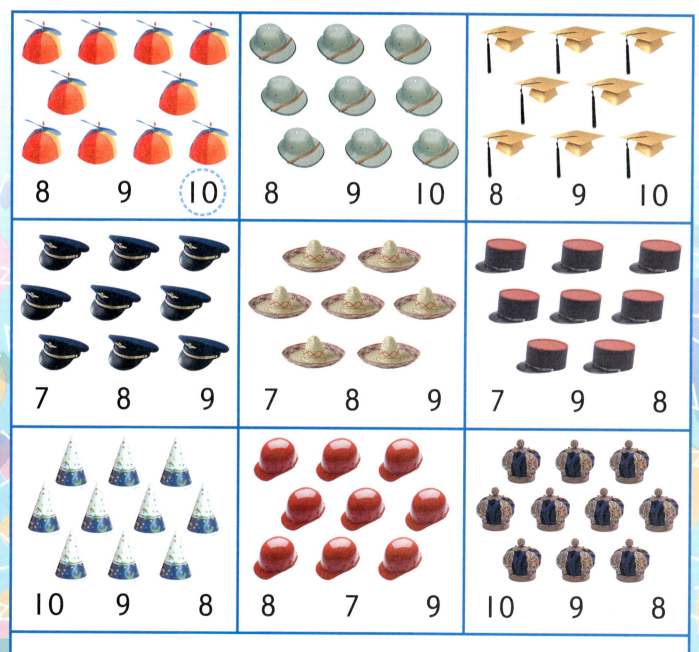

Draw fewer than 10 .

8 to 10

five 5

_____ can count pennies.

1 penny  2 pennies  3 pennies

1 cent 1¢   2 cents 2¢   3 cents 3¢

Write how much.

1.   4 ¢

2.   ___ ¢

3.   ___ ¢

4.
     ___ ¢

5.   ___ ¢

6.
     ___ ¢

7. Show more than 5 pennies in your Math Journal.

Count by 1: Pennies

_____ can join.

How many 🚢 in all?

Show 4. 🚢🚢🚢🚢

Show 2. 🚢🚢

　　　6 in all.

Join to find how many in all.

Join. Use ■. Write how many in all.

1. ____ in all

2. ____ in all

3. ____ in all

4. ____ in all

Draw ▲ to join. Write how many in all.

5. Show 4.
   Show 1.

   ____ ▲ in all

6. Show 1.
   Show 4.

   ____ ▲ in all

Join

seven 7

_____ can separate.

How many are left?

Show 4.
Take away 1.
3 are left.

Take away to **separate**.

Separate. Use ■. Tell how many you take away. Write how many are left.

1.

   ____ are left.

2.

   ____ are left.

3.

   ____ are left.

4.

   ____ is left.

Draw ●. Take away. Write how many are left.

5. Show 5.
   Take away 2.

   ____ are left.

6. Show 5.
   Take away 3.

   ____ are left.

8 eight   **Separate**

_____ can identify positions.

 1. Draw a 🪣 next to the 🪣.

2. Draw a ☁️ up in the sky.

Ring **Yes** or **No**.

3. Is the 🕊 **on top of** the ⛱?   (Yes)   No

4. Is the 🪑 **under** the ⛱?   Yes   No

5. Is the 🦀 **in front of** the 🪣?   Yes   No

6. Is the 🪑 **far** from the 🔺?   Yes   No

7. Is the 🔺 **on the right**?   Yes   No

8. Is the 🧒 **above** the 🪁?   Yes   No

9. Is the 🪁 **down** on the ground?   Yes   No

Positions

_____ can read a calendar.

Each day in 1 week I read a story.

Sunday   Monday   Tuesday   Wednesday   Thursday   Friday   Saturday

1. There are _____ days in 1 week.

2. _____ is the first day of the week.

3. The last day of the week is _____.

I can make a calendar for the month of _____.

Sunday	Monday	Tuesday	Wednesday	Thursday	Friday	Saturday

4. This month begins on _____.

5. This month ends on _____.

Calendar

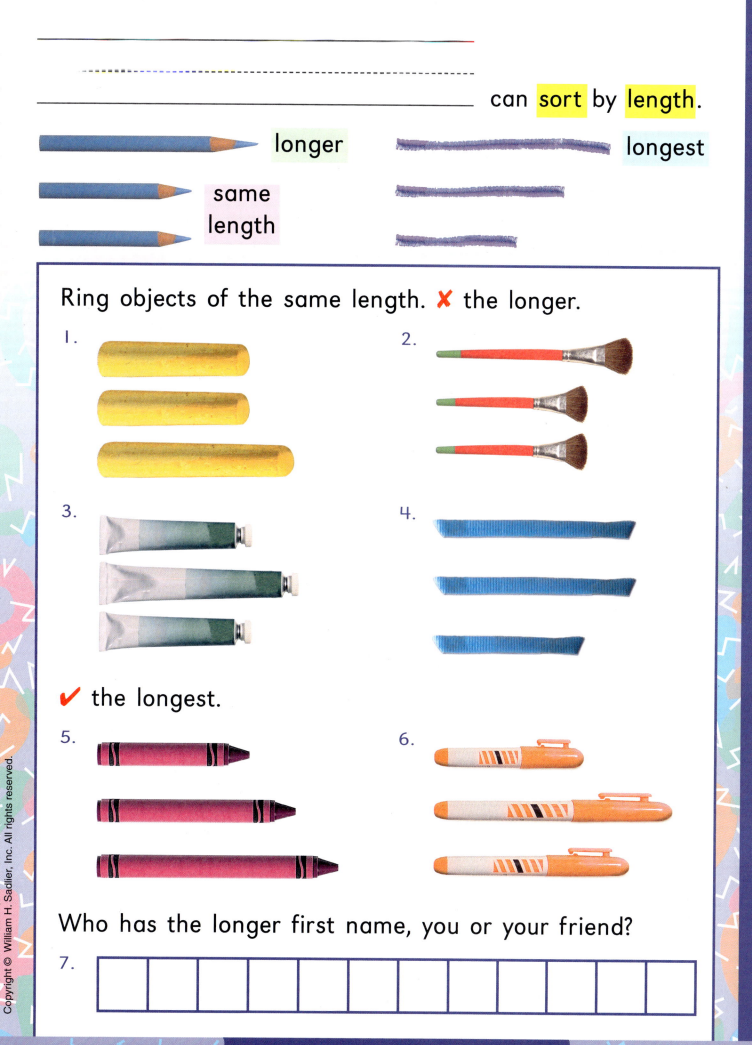

_____

_____ can sort by length.

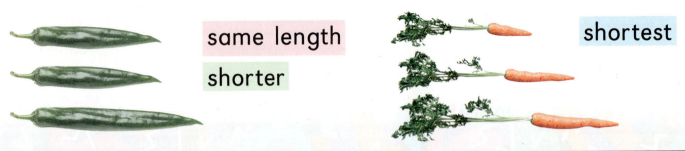

same length
shorter
shortest

Ring objects of the same length. ✔ the shorter.

1.    2.

✔ the shortest. ✘ the longest.

3.    4.

Look at this pattern.
Draw a shorter pattern.

5.

6. What is the opposite of shorter? _____

REVIEW OF GRADE K SKILLS

12 twelve **Same, Shorter, Shortest**

Introduction to Problem Solving

This magazine belongs to
_____

To be a super problem solver, use these steps.

Read → **Draw** → **Think** → **Write** → **Check**

- **Read**: Put yourself in the problem.
- **Draw**: Picture what is happening. Study the facts. Know what the question asks.
- **Think**: Plan what you will do to solve the problem.
- **Write**: Work your plan. Ring your choice. Sometimes you will add or subtract.
- **Check**: List and label your answer. Be sure it makes sense.

Use this strategy: Draw a Picture.

Read: There are 3 . First 1 flies away. Then 2 fly away. Are there any left?

Draw: Draw a picture of 3 spaceships. ✗ out each that flies away.

Think: Subtract to tell what happens.

$$3 - 1 - 2 = ?$$

Write:

Check: Use 🟩 to act it out.

INTRODUCTION TO PROBLEM SOLVING

Here are some problem-solving strategies.

- Act It Out
- Draw a Picture
- Find a Pattern
- Write a Number Sentence
- Choose the Operation
- Draw to Compare
- Use a Model
- Use Logical Reasoning
- Make a Table
- Use a Graph
- Ask a Question
- Find Hidden Facts
- Use a Map
- Guess and Test
- Find Extra Information

Read → **Draw** → **Think** → **Write** → **Check**

Use this strategy: **Act It Out**

Read
You see 1.
Next you see 2.
Last, you see 3 more.
How many martians do you see in all?

Draw
Draw a picture.
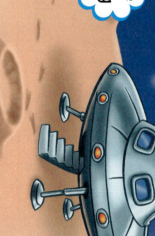

Think
Use  to act it out.
Count the  to find how many in all.

Write

___, ___, ___, ___, ___, ___

Check
So 1 and 2 and 3 equals ___ martians in all.

Numbers to 12

CRITICAL THINKING

Leo sorted the bugs by how they move. Name 1 other way you can sort the bugs.

Math Alive at Home

For more information about Chapter 1, visit the Family Information Center at **www.sadlier-oxford.com**

Dear Family,

Today your child began Chapter 1. As she/he studies numbers to 12, you may want to read the poem below, which was read in class. Have your child talk about some of the math ideas shown on page 15.

Look for the 🏠 at the bottom of each skills lesson. Use the suggestion to help your child improve in math. You may want to have pennies and countables available for your child to use throughout this chapter.

Home Reading Connection

Miss Spider's Tea Party

In this story, Miss Spider is sad because no insects will come to her tea party. The insects are afraid that they will be the spider's meal. Then, one day, everything changed....

Ten tiny steaming cups of tea
Were perched upon her trembling knee.
She sipped and sobbed, and heard a cough
And turned to see a small wet moth–
A fragile thing so soaked by rain,
His wings too damp to fly again.

She smiled and took a checkered cloth
To cloak the frail and thankful moth.
They talked and snacked on tea and pie
Until his tiny wings were dry.
Then lifting him with tender care
She tossed him gently in the air.

The moth told Ike, then Ike told May,
Who went from bug to bug to say,
"There is no reason for alarm.
She's never meant us any harm!"
So later on that afternoon,
Assembled in the dining room,
Eleven insects came to tea
To share Miss Spider's courtesy.

David Kirk

Home Activity

Number Pick

Try this activity with your child. Make a set of number cards for 0–12. As your child learns each number, place that number card in a bag. Have your child draw a number from the bag, write the number, name objects that come in that number, and draw a picture to show some ways she/he can arrange a pattern.

Copyright © William H. Sadlier, Inc. All rights reserved.

Name _____

MATH CONNECTIONS
Real World

Write your telephone number.

___ ___ ___ — ___ ___ ___ ___

Help Queen Bee call Miss Ladybug. Use the numbers below to make up a telephone number.

___ ___ ___ — ___ ___ ___ ___

You can put this in your Math Portfolio.

seventeen **17**

CROSS-CURRICULAR CONNECTIONS
Language Arts

Name _____

This is a **tally**.

Tally	I	II	III	IIII	IIII̶	IIII̶ I
Number	1	2	3	4	5	6

 Use the maze below to tally the **data**.

1. Go to Start.
   Go 2 🍃 down.
   Move 4 🍃 to the right.
   Tally the 🐞. IIII
   Write the number. _____

2. Go to Start.
   Go 1 🍃 down.
   Move 3 🍃 to the right.
   Tally the 🐝. _____
   Write the number. _____

3. Go to Start.
   Go 2 🍃 down.
   Move 2 🍃 to the right.
   Go 1 🍃 down.
   Tally the 🐜. _____
   Write the number. _____

4. Go to Start.
   Go 1 🍃 down.
   Move 5 🍃 to the right.
   Go 2 🍃 down.
   Tally the 🪲. _____
   Write the number. _____

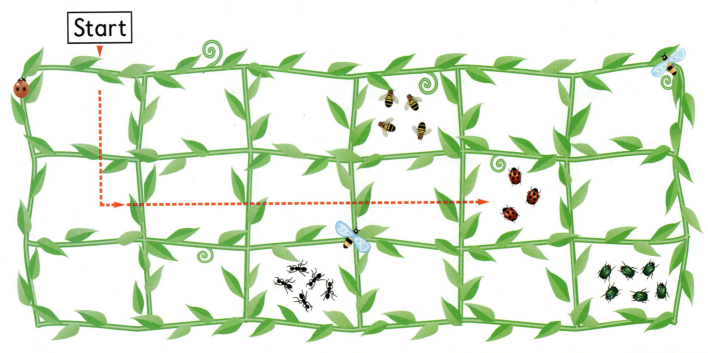

18 eighteen

You can put this in your Math Portfolio.

Name _____

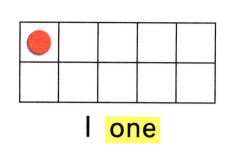

1 one

Write One and Two

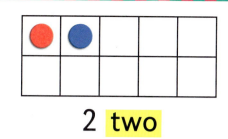

2 two

Begin at the dot. Write the number.

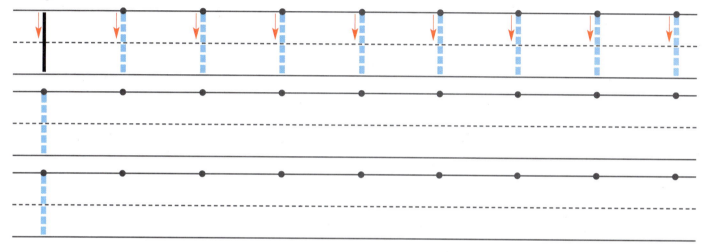

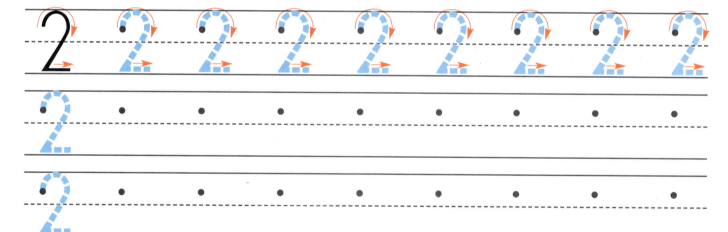

Color one  red . Color two  blue .

 How many 🐛 are not colored?

Have your child count and identify groups of 1 and 2 objects.

nineteen **19**

Write the number word and the number.

Write the number word and the number.
Color that many.

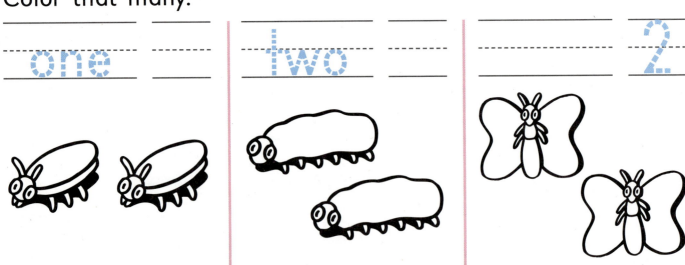

 Draw a picture with 1 bug.
Draw a picture with 2 bugs.

Name _____

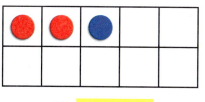

3 three

Write Three and Four

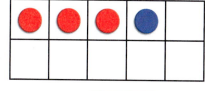

4 four

Begin at the dot. Write the number.

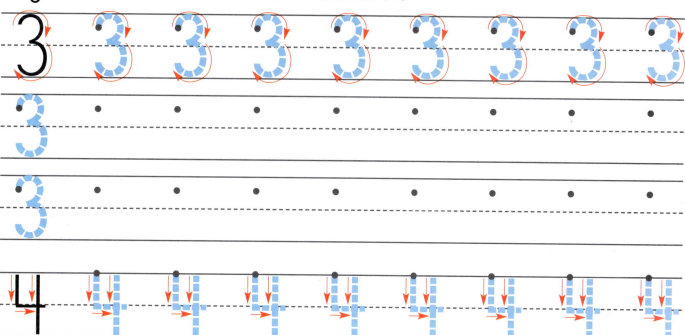

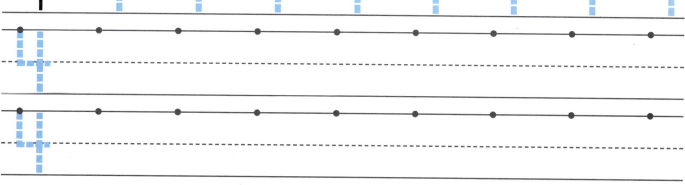

How many 🌸 are left?

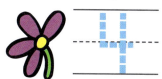         

 Describe the pattern pictured above.

Write words of 3 or 4 letters—for example: *cat* and *tree*. Ask your child to count the letters in each word.

twenty-one **21**

Write the number word and the number.

three   four   three

four   three   four

Write the number word and the number.
Color that many.

three   four   _____ 3

PROBLEM SOLVING Ring groups less than 4.

  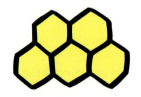 

Name _____

Write Five and Zero

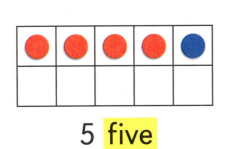

5 five

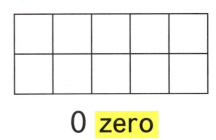

0 zero

Begin at the dot. Write the number.

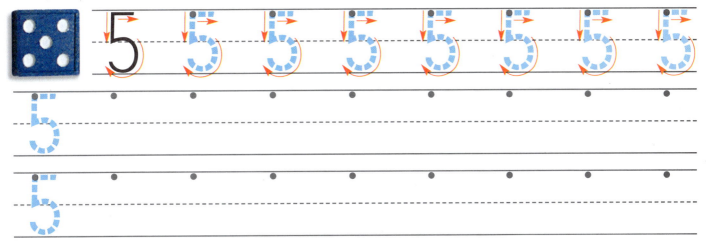

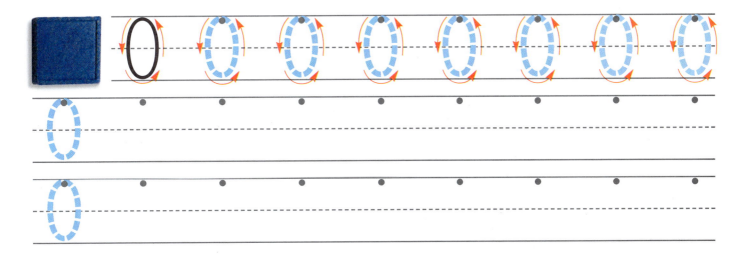

Write how many • on each part.

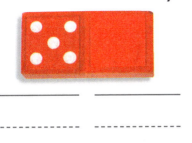

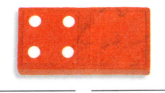

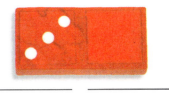

Put 5 or fewer objects in one of your hands.
Have your child guess which hand has 0 objects.

twenty-three **23**

Write the number word and the number.

Write the number word and the number.
Draw ● for each number.

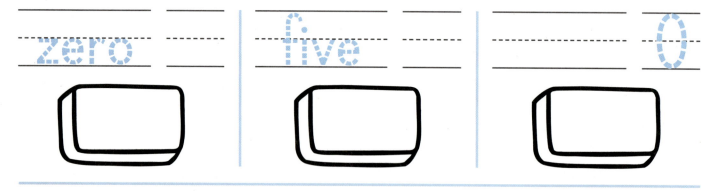

 Listen to the directions.

 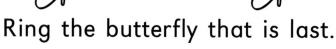 

Ring the butterfly that is last.

Name _____

One More, One Fewer

3 is one fewer than 4.

5 is one more than 4.

Ring how many in the group.	Draw ■ to show one more.	Draw ■ to show one fewer.

1.
   3   ④   5

2. (3 yellow squares)
   1   2   3

3. (2 green squares)
   2   3   4

4. (1 yellow square)
   0   1   2

1-4 Ask your child to use countables to show one more than 4 and 1 fewer than 4. Repeat with the numbers 3 and 2.

twenty-five **25**

Listen to the directions.

Ring how many ●.

1   2   ③   4   5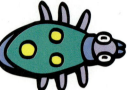

Draw one fewer ●.

   2

Draw one more ●.

Ring how many ●.

1   2   3   4   5

Draw one fewer ●.

Draw one more ●.

Ring how many ∞.

1   2   3   4   5

Draw one fewer ∞.

Draw one more ∞.

Ring how many ◡.

1   2   3   4   5

Draw one fewer ◡.

Draw one more ◡.

26   twenty-six

Name _____

Picture Graph and Pictograph

A **picture graph** uses pictures to show how many.

Sort the bugs to make a picture graph.

Color 1 picture for each bug. Then write the number.

one	🕷 🕷 🕷 🕷 🕷	
two	🦋 🦋 🦋 🦋 🦋	
three	🐞 🐞 🐞 🐞 🐞	
four	🪲 🪲 🪲 🪲 🪲	
five	🐝 🐝 🐝 🐝 🐝	

 Use the picture graph above. Ring the bug that is:

1. one more than .

2. one fewer than .

3. There are ____ more than .

Have your child use the data pictured to make up a question for you to answer.

twenty-seven **27**

Use these tallies to make a **pictograph**.

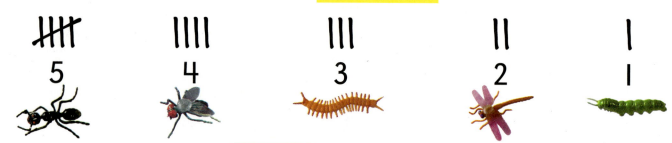

A pictograph uses a **symbol** to show how many.

Draw 1 ☺ for each **tally mark**.

My Bug Collection

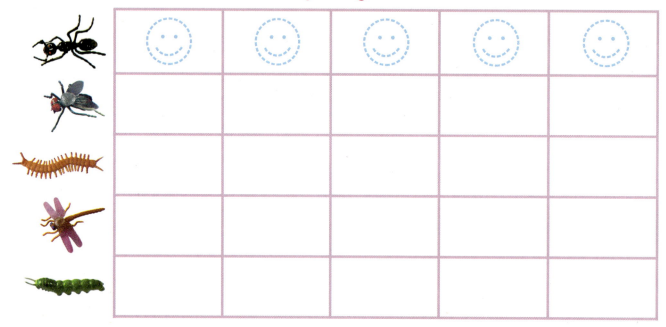

Listen to the directions to **survey** your classmates.

Favorite Color

Name _____

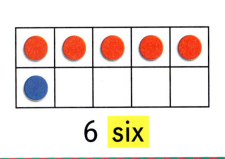

6 six

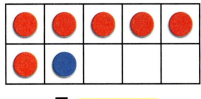

7 seven

Begin at the dot. Write the number.

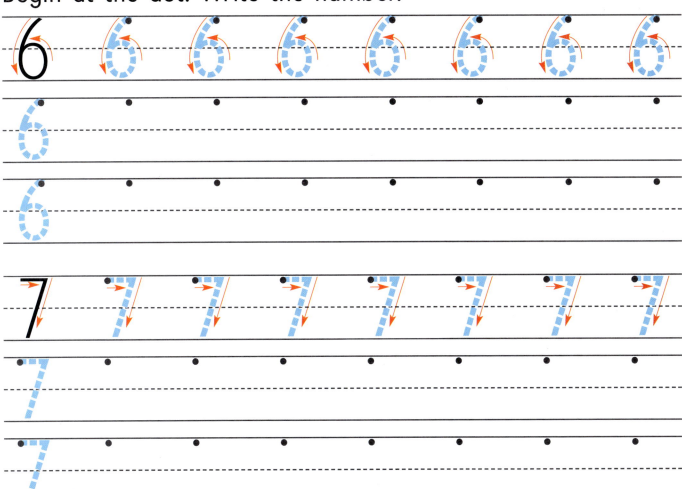

Draw 1 more bug. Then write how many.

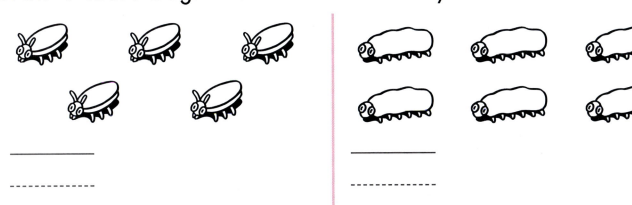

1-6 Show your child a set of 5 objects. Ask her/him how to make it a set of 6 and then a set of 7.

twenty-nine **29**

Write the number word and the number.

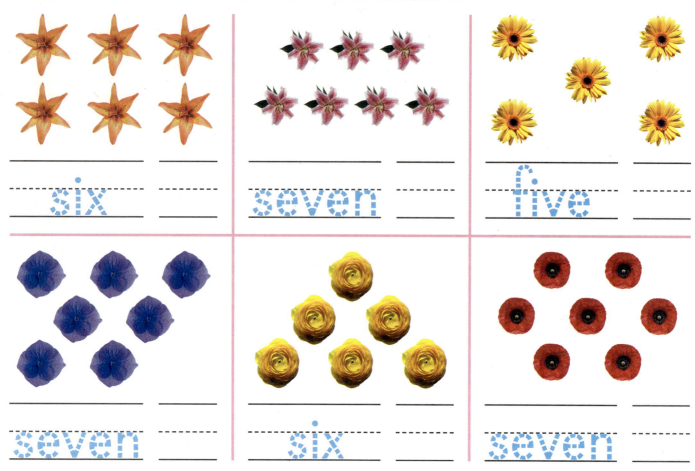

Write the number word and the number.
Color only that many.

 CHALLENGE

Draw 6 O in all. Put the same number on each side.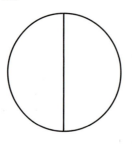

Draw 7 O in all. Put 1 more O on the left side.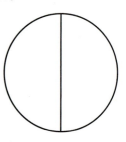

30 thirty

Name _____

8 eight

Write Eight and Nine

9 nine

Begin at the dot. Write the number.

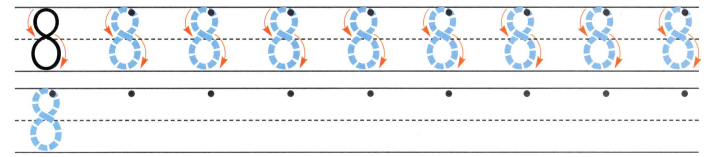

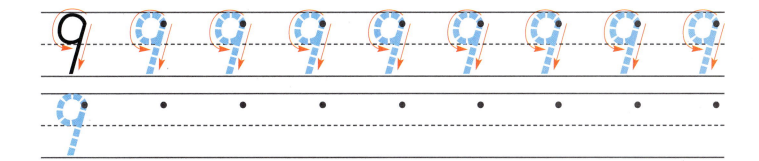

Write the number that is one fewer.

1 fewer than ▨ is _____.

1 fewer than ▨ is _____.

1 fewer than ▨ is _____.

Have your child draw dominoes to show 1 more than 5, 6, 7, and 8.

thirty-one 31

Write the number word and the number.

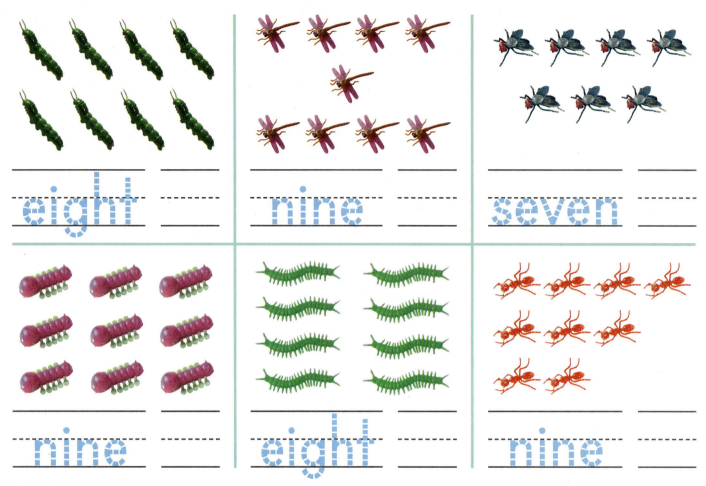

Write the number word and the number.
Color only that many.

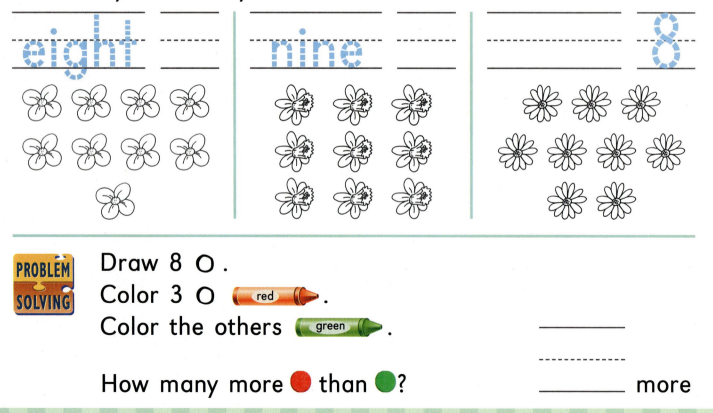

Draw 8 ○.
Color 3 ○ red.
Color the others green.

How many more ● than ●?

_____ more

Name _____

Write Ten

10
ten

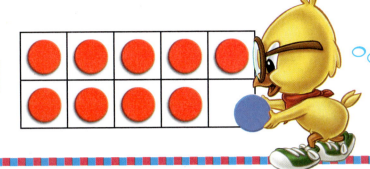

10 counters fill the ten-frame.

Begin at the dot. Write the number.

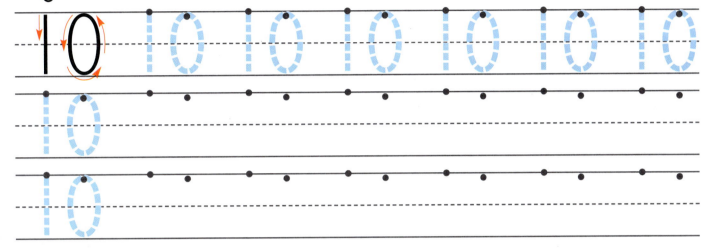

Write the number word and the number.

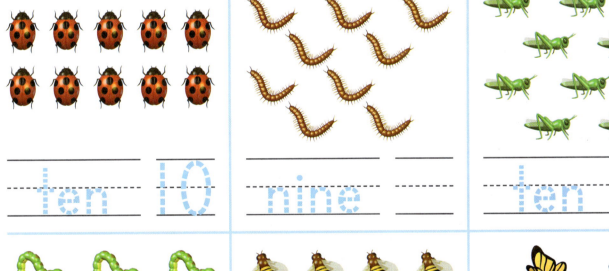

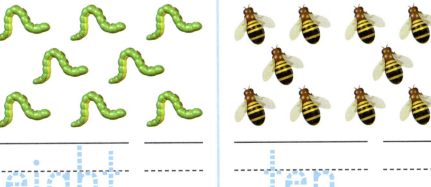

Have your child trace her/his hands and number each finger.

thirty-three 33

Write how much each bug costs to make.

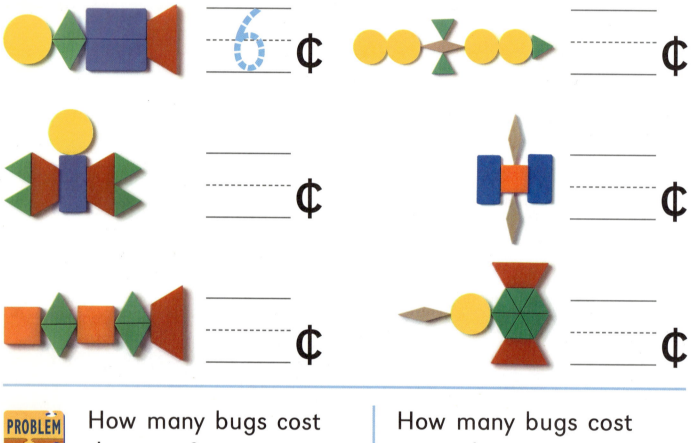

 How many bugs cost the same? _____

How many bugs cost more than 6¢? _____

 Draw a bug that costs 10¢.

Draw a bug that costs 5¢.

Name _____

Write Eleven and Twelve

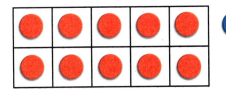

11 eleven

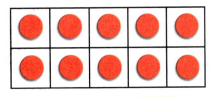

12 twelve

 11 is how many more than 10?
12 is how many more than 10?

Begin at the dot. Write the number.

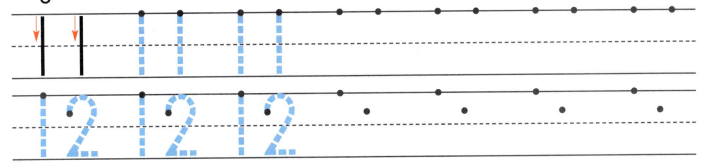

Write the number word and the number.
Draw 🔴 to show the number.

 _____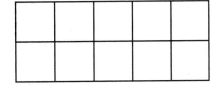

twelve _____

eleven _____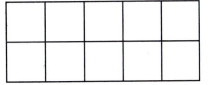

12 🟩 are 1 dozen.

Ring a dozen 🟩.

Are there more than
1 dozen 🟩? _____

How many more? _____ more

Have your child make patterns for 11 and 12 using small countables.

thirty-five **35**

"2 more make 11."

11 🪙 is 2 more than 9 🪙.

Use 🪙 and a ▦.
Write the missing numbers.

1. 10 🪙 is ___ more than 9 🪙.

2. 10 🪙 is ___ more than 8 🪙.

3. 10 🪙 is ___ more than 7 🪙.

4. 10 🪙 is ___ more than 6 🪙.

5. 10 🪙 is ___ more than 5 🪙.

6. 10 🪙 is ___ more than 4 🪙.

TALK IT OVER Tell the pattern you see in 1–6.

MATH JOURNAL 7. Write the next three sentences.

36 thirty-six

Name _____

Zero to Twelve

Draw to show each number.
Color to complete the pattern.

zero	0
one	1 ●
two	2 ● ■
three	3 ● ■ ●
four	4 ● ■ ● ■
five	5 ● ■ ● ■ ●
six	6 ○ □ ○ □ ○ □
seven	7 ○ □ ○ □ ○ □ ○
eight	8 ○ □ ○ □ ○ □ ○ □
nine	9 ○ □ ○ □
ten	10 ○ □ ○
eleven	11 ○ □
twelve	12

 Look at the drawings above.
Describe the patterns you see.

1-10 Place 12 small countables in a bag. Have your child grab a handful. Have him/her tell how many and write the word and number.

thirty-seven **37**

Draw each group. Write the number.

1. twelve 🍎 and ten 🍏

   ___ 🍎

   ___ 🍏

2. seven 🐛 and four 🐛

   ___ 🐛

   ___ 🐛

3. eight 🌷 and five 🌷

   ___ 🌷

   ___ 🌷

4. three 🦋 and six 🦋

   ___ 🦋

   ___ 🦋

SECOND LOOK In 1–4 ✔ more and ✘ fewer.

TALK IT OVER How can you make the groups in 1–4 have the same number?

CRITICAL THINKING 5. Ring each group.

between five and nine

between nine and twelve

38 thirty-eight

Name _____

Order 0–12

Numbers have a special <mark>order</mark>.

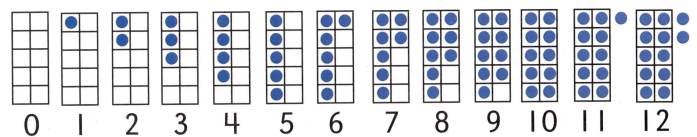

 Tell how a ten-frame helps you order numbers.

Write the missing numbers.

0, __1__, 2, ___, 4, ___, 6, ___, 8, ___, 10, ___, 12

0, 1, 2, ___, 4, 5, ___, 7, 8, ___, 10, 11, ___

___, 1, 2, 3, ___, 5, 6, 7, ___, 9, 10, 11, ___

0, 1, ___, ___, 4, 5, ___, ___, 8, 9, ___, ___, 12

Write how many.

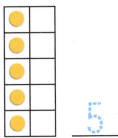 __5__   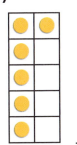 ___   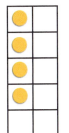 ___   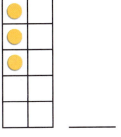 ___

Now order the numbers. ___ ___ ___ ___

Have your child order a set of 0–12 number cards.

thirty-nine **39**

This is a **number line**. Describe it.

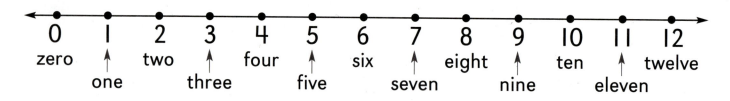

Write the missing numbers. Use the number line.

1. 2, _3_, ___, 5
2. ___, 8, ___, 10
3. ___, 1, 2, ___
4. 5, ___, 7, ___
5. 9, ___, ___, 12
6. ___, 4, ___, 6

Write the missing number word.

7. eight, _____, ten
8. four, five, _____

9. Connect the dots in order.

40 forty

Name _____

Counting on a Number Line

Count on in order.

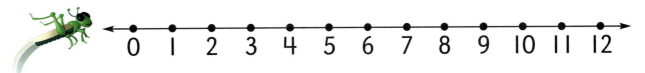

Write the missing numbers.

1.
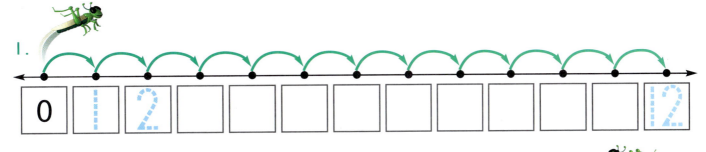

2.
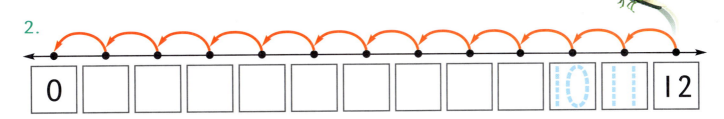

3.
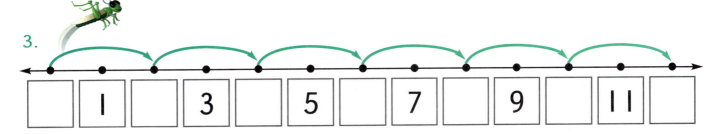

4.
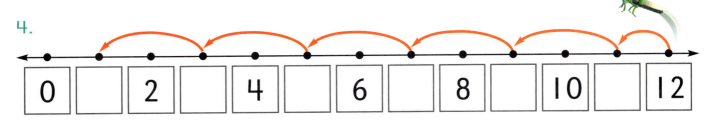

5.
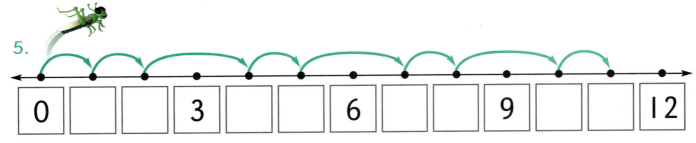

Listen. Color the boxes. Talk about the pattern.

Ask your child to describe the number line pattern she/he likes best.

forty-one **41**

Hopper counts on or back to reach a place.
Write the number of hops.

1.

   __4__ hops to the 🌻. ____ more hops to the 🌾.

2.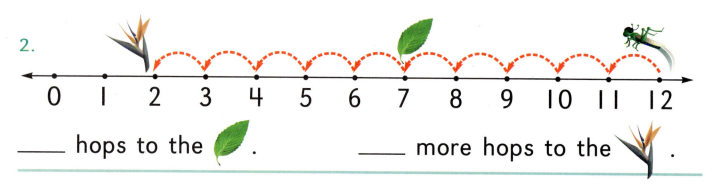

   ____ hops to the 🍃. ____ more hops to the 🌸.

Draw the hops. Write the number.

3.

   ____ hops from 🦗 to the 🌸.

   ____ hops from the 🌸 to the 🌻.

 Choose a number on the ⟵⟶. Hop left from 🦗.
Draw the hops. Complete.

4.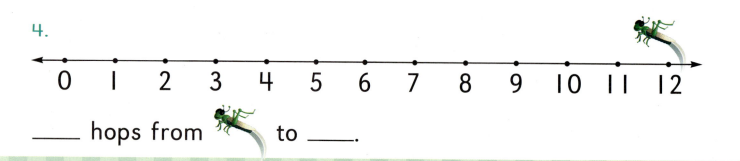

   ____ hops from 🦗 to ____.

Name _____

Counting Back

The ants start at 12.
Count back by 1.
Write the hidden numbers. 8 ___ ___ ___

Count back by 1. Write the missing numbers.

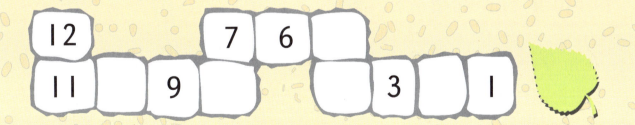

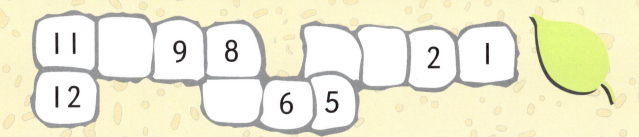

Count back from any number from 12 or less, and ask your child to continue to zero.

forty-three 43

Before, After, Between

Name _____

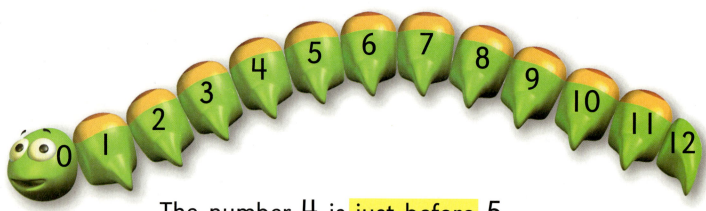

The number 4 is just before 5.
The number 6 is just after 5.
The number 5 is between 4 and 6.

What number comes just before?

_6_, 7   ___, 2   ___, 9   ___, 11

___, 12   ___, 8   ___, 10   ___, 4

What number comes just after?

3, _4_   8, ___   9, ___   1, ___

0, ___   11, ___   6, ___   7, ___

What number comes between?

5, _6_, 7   2, ___, 4   8, ___, 10

7, ___, 9   10, ___, 12   4, ___, 6

0, ___, 2   9, ___, 11   3, ___, 5

Use the words before, between, and after to tell about any number from 1 to 11.

44 forty-four

Ask your child to choose a number from 1–11. Have him/her describe it—for example, 8 is before 9, after 7, between 7 and 9.

1-14

Name _____

Compare

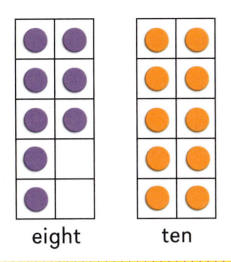

eight   ten

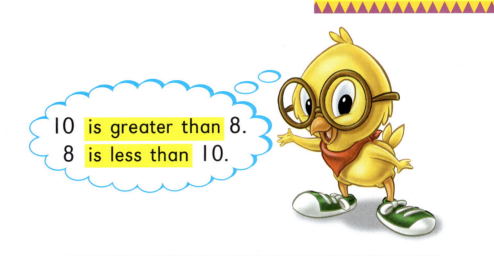

10 is greater than 8.
8 is less than 10.

Use a ▦ and ◯. Model and compare.

five	seven

__7__ is greater than __5__.

____ is less than ____.

twelve	nine

____ is greater than ____.

____ is less than ____.

ten	six

____ is greater than ____.

____ is less than ____.

four	eight

____ is greater than ____.

____ is less than ____.

 Color the numbers greater than 7 .

 9   4   8   6   10

forty-five **45**

Odd and Even

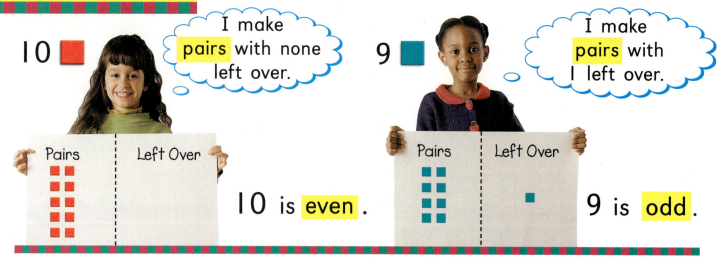

Draw to make pairs and leftovers.
Ring odd or even.

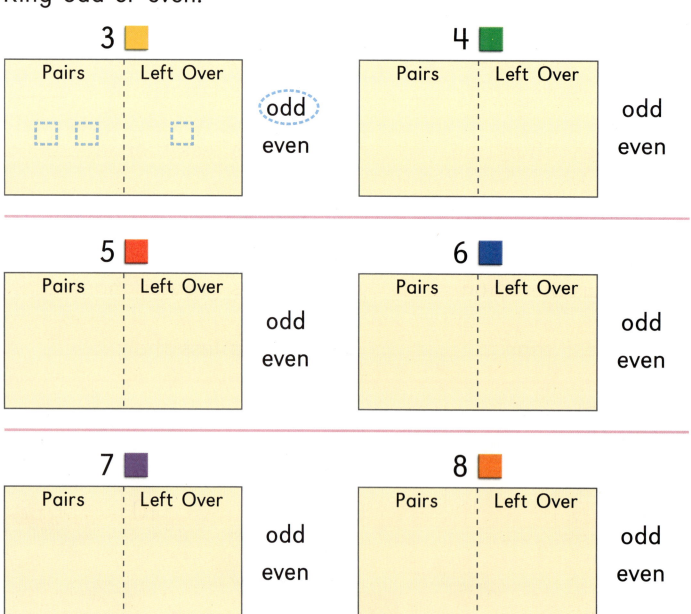

Name _____

Ordinals

🔊 Listen to the directions. Color each birdhouse.

first　　second　　third　　fourth　　fifth

Write the place of the birdhouse.

1. 　　first

2. 　　second

3. 　　_____

4. 　　_____

5. 　　_____

🔊 6. Listen to the directions. Color each bird.

Have your child describe the position of 5 objects using ordinal numbers.

forty-seven **47**

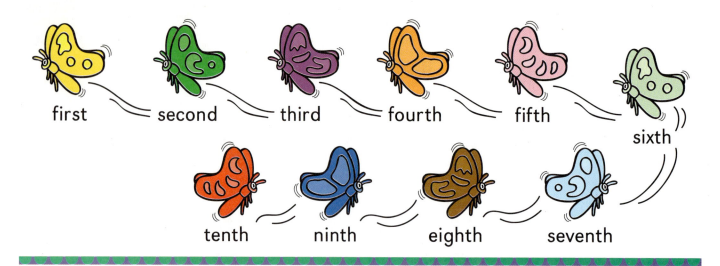

 Listen to the directions.
Ring the correct place of each butterfly.

third
(fourth)
fifth

eighth
ninth
tenth

seventh
eighth
sixth

first
fourth
seventh

ninth
second
tenth

sixth
third
first

seventh
tenth
sixth

eighth
seventh
first

third
fourth
second

48 forty-eight

Name _____

PROBLEM-SOLVING STRATEGY
Find a Pattern

Read → Think → Write

What comes next in each pattern?

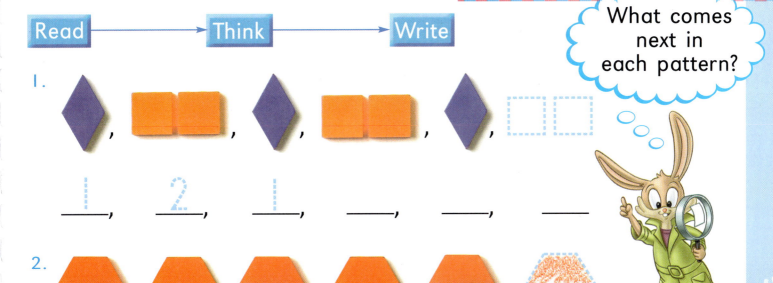

1. ___, ___, ___, ___, ___, ___

2. ___, ___, ___, ___, ___, ___

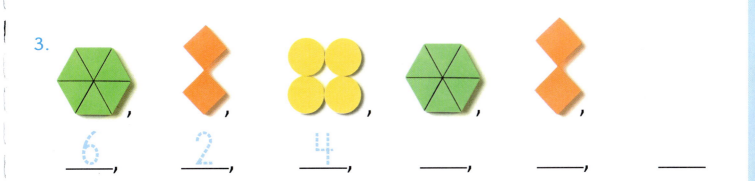

3. ___, ___, ___, ___, ___, ___

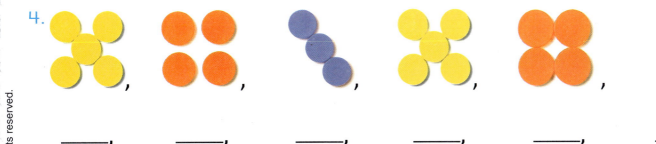

4. ___, ___, ___, ___, ___, ___

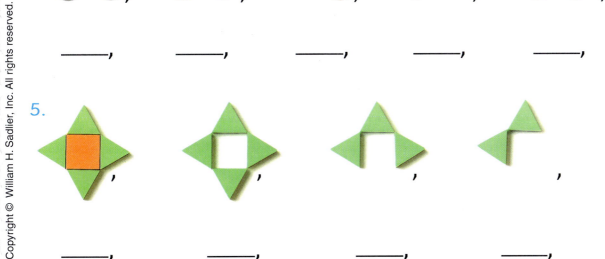

5. ___, ___, ___, ___

___, ___, ___, ___

Ask your child to tell how the patterns in exercises 1, 4, and 5 differ. Have him/her draw what would be next in each.

forty-nine **49**

Read → Think → Write

Use  to model each number. Write what comes next.

1. 6, 8, 6, 8, _6_, ___

2. 10, 11, 12, 10, 11, 12, ___, ___, ___

3. 8, 9, 9, 8, 9, 9, ___, ___, ___

4. 6, 6, 7, 6, 6, 7, ___, ___, ___

5. 6, 3, 0, 6, 3, 0, ___, ___, ___

6. 7, 9, 11, 7, 9, 11, ___, ___, ___

7. 1, 3, 5, 7, 1, 3, 5, 7, ___, ___

 Make up two patterns.

8. Use shapes. _____

9. Use numbers. _____

Name _____

PROBLEM-SOLVING APPLICATIONS
Draw a Picture

Read ———————→ Draw ———————→ Write

1. has 3 .

    has 1 more 🍃 than 👧.

   How many 🍃 does 👧 have?

    has 4 🍃.

2. draws 3 .

    draws 1 fewer 🦋.

   How many 🦋 does 👦 draw?

   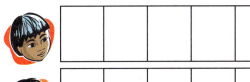

    draws ___ 🦋.

3. makes 4 .

    makes as many as .

   How many 🌼 does 👦 make?

   👦 makes ___ 🌼.

4. sees 5 🐛.

    sees 1 fewer .

   How many 🐛 does 👧 see?

   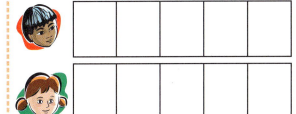

    sees ___ 🐛.

Ask your child to explain how she/he solved one of these problems by drawing a picture.

fifty-one 51

Listen to the directions.

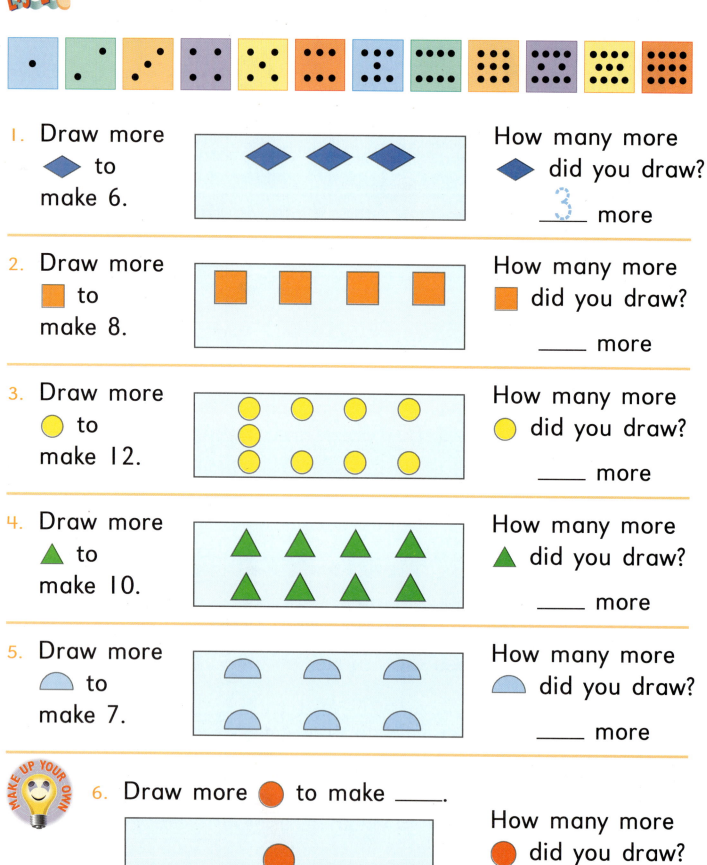

Name _____

Chapter Review and Practice

Tally the spots on each. Write the number.

	Tally	Number
1. 	‖‖‖‖	_____
2.		_____
3.		_____
4.		_____

5. Draw dots to show 1 fewer than 5.

6. Draw dots to show 1 more than 4.

Write the number. Use a ↔.

6 is just before _____.

9 is just after _____.

10 is just after _____.

4 is just before _____.

11 comes between _____ and _____.

1 comes between _____ and _____.

This page reviews the mathematical content presented in Chapter 1.

fifty-three 53

Name _____

Performance Assessment

1. Pick one number card.

   | 2 | 4 | 6 | 8 |

   Ring yes or no.

   Is the number even?   yes   no
   Is the number odd?   yes   no
   Is it less than 10?   yes   no

   Draw dots to show a group with 1 fewer and a group with 1 more than the number.

   1 fewer   1 more

Draw the hops. Write the number.

2.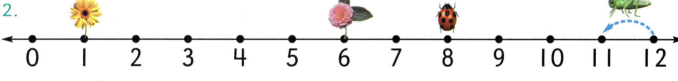

   ____ hops to the 🐞.   ____ hops from  to 🌻.

Choose one of these projects.
Use a separate sheet of paper.

3. Draw and color 10 bugs in a row. Write about each.

   The first bug is red.

4. Pick a number from 0 to 12. Draw different groups to show the number.

   Ways to show twelve

fifty-five **55**

Check Your Mastery

Name _____

Write the missing numbers.

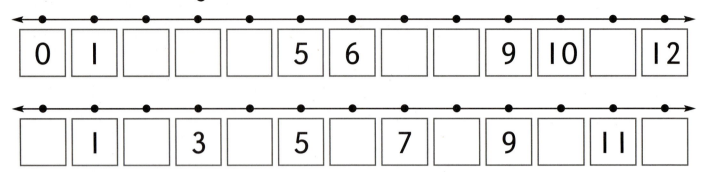

✔ the seventh shape. ✘ the ninth shape.

Draw and color each shape.   eighth    fourth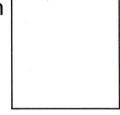

Write two numbers greater than 7. Draw each group.

Write two numbers less than 5. Draw each group.

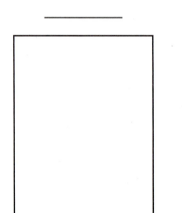 

56 fifty-six

This page is a formal assessment of your child's understanding of the content presented in Chapter 1.

Understanding Addition Facts to 6

2

CRITICAL THINKING

Sara bought 1 fish. Now she has 5 fish altogether. How many fish did she have at first? Draw and color the fish she has now.

57

Math Alive at Home

Dear Family,

Today your child began Chapter 2. As your child studies addition facts to 6, you may want to read the story below, which was read in class. Have your child talk about some of the math ideas pictured on page 57.

Look for the 🏠 at the bottom of each skills lesson. The suggestion on the page gives you an opportunity to improve your child's understanding of math and to reinforce her/his math language. You may want to have pennies and small countables available for your child to use throughout this chapter.

For more information about Chapter 2, visit the Family Information Center at www.sadlier-oxford.com

Home Activity

Top-Hat Sums

Try this activity with your child. Draw an outline of a top hat on a sheet of construction paper (see diagram). As your child completes this chapter, learning each sum to 6, have him/her list all the facts for that sum on a hat. At the end of the chapter, when your child has completed all the hats, ask her/him to explain why the list of number sentences became longer as the sums became larger.

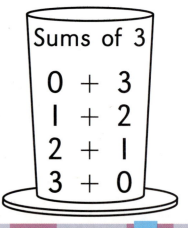

Home Reading Connection

On this night when there's a storm.
A hat can surely keep you warm.
So after the pet store is closed up tight.
You'll be surprised to learn what pets do at night.

Hats

A hat for a hamster,
A hat for a dog,
A hat for a goldfish,
A hat for a frog,
A hat for me
To wear in cold weather,
How many hats
Have we got altogether?

Daphne Lister

Name _____

MATH CONNECTIONS
Modeling Figures

Listen to the story.

Write how many in all of each color.

1. _____   2. _____   3. _____

4. needs _____  to make.

5. Show how many ▢ you use to cover each.

CONNECTIONS

You can put this in your Math Portfolio.

fifty-nine **59**

CROSS-CURRICULAR CONNECTIONS
Science

Name _____

Sort the pet store animals.

1. How many legs does each pet have?

     4   ____    ____    ____

2. Where does each live? Match.

 Use the clues to match.

3. I have more than 3 legs.
   I eat 🦴 .

4. I live in water.
   I have 0 legs.
   I swim.

5. I have less than 3 legs.
   I fly.

6. Follow directions. Use  .

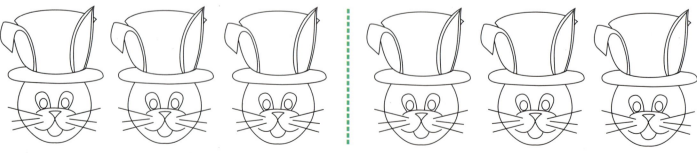

60 sixty

You can put this in your Math Portfolio.

Name _____

Understanding Addition

Let's put our tiles together. I have 2 ■.

I have 3 ■. How many tiles do we have altogether?

2 and 3 equals 5 in all.

Color the tiles. Tell how many altogether.

1.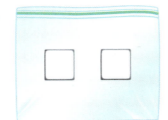

   __2__ ■ and __2__ ■

   equals ____ in all.

2.

   ____ ■ and ____ ■

   equals ____ in all.

3.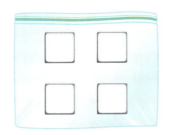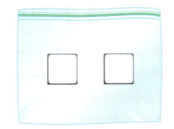

   ____ ■ and ____ ■

   equals ____ altogether.

Put a total of 6 or fewer countable objects into two clear plastic bags. Ask your child to tell you the addition story.

sixty-one **61**

Use 🪙. Write how many cents in all.

1.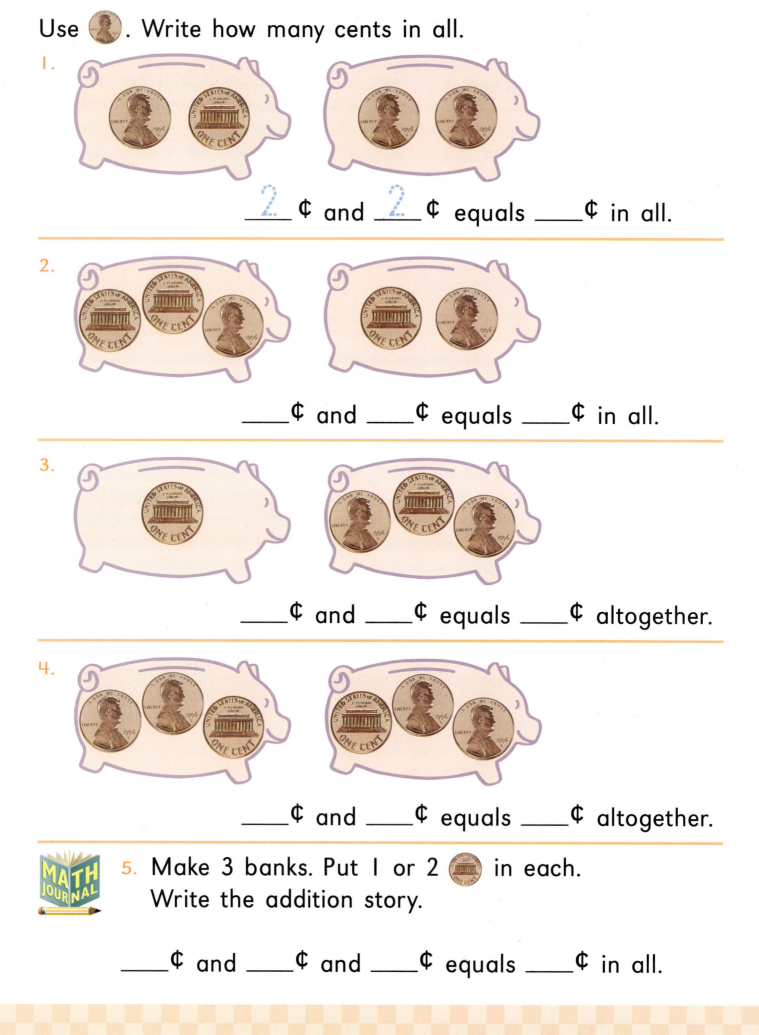

____¢ and ____¢ equals ____¢ in all.

2. ____¢ and ____¢ equals ____¢ in all.

3. ____¢ and ____¢ equals ____¢ altogether.

4. ____¢ and ____¢ equals ____¢ altogether.

5. Make 3 banks. Put 1 or 2 🪙 in each. Write the addition story.

____¢ and ____¢ and ____¢ equals ____¢ in all.

Name _____

Addition Sentences

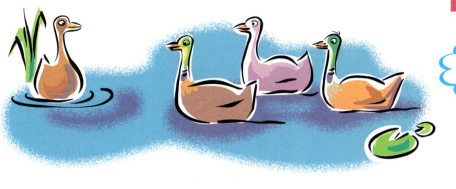

1 + 3 = 4 is a **number sentence**.

1 + 3 = 4
plus **equals**

Add.

1.

   2 + 1 = 3

2.

   1 + 1 = ___

3.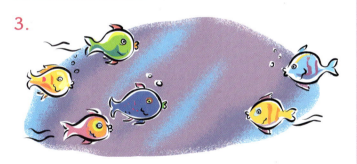

   4 + 2 = ___

4.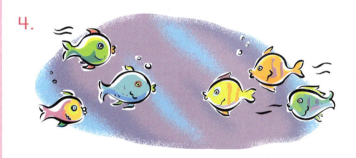

   3 + 3 = ___

5.

   1 + 4 = ___

6.

   2 + 2 = ___

Have your child use 5 pennies to tell an addition story. Then have her/him write an addition sentence for the story using the symbols + and =.

sixty-three **63**

Add. Fill in each addition sentence.

1.

5 + 1 = 6

2.

___ + ___ = ___

3.

___ + ___ = ___

4.

___ + ___ = ___

5.

___ + ___ = ___

6.

___ + ___ = ___

7.

___ + ___ = ___

8.

___ + ___ = ___

Name _____

Sums of 4

 Listen to the **addition story**.
You can write **addition facts** in 2 ways.

$3 + 1 = \underline{}$

$$\begin{array}{r} 3 \\ +1 \\ \hline 4 \end{array}$$ addend
addend
sum

Add.

1.  $\begin{array}{r} 2 \\ +2 \\ \hline 4 \end{array}$

2.  $\begin{array}{r} 1 \\ +3 \\ \hline \end{array}$

3.  $\begin{array}{r} 0 \\ +4 \\ \hline \end{array}$

4.  $\begin{array}{r} 3 \\ +1 \\ \hline \end{array}$

5.  $\begin{array}{r} 1 \\ +2 \\ \hline \end{array}$

6.  $\begin{array}{r} 1 \\ +1 \\ \hline \end{array}$

7.  $\begin{array}{r} 4 \\ +0 \\ \hline \end{array}$

8.  $\begin{array}{r} 2 \\ +1 \\ \hline \end{array}$

SECOND LOOK In 1–8 ring sums of 4.

 Name the addends in each fact above.

 Have your child join objects to show the sums of 4.

sixty-five **65**

Find the sum. Use your ■.

1. $\begin{array}{r}2\\+2\\\hline 4\end{array}$ 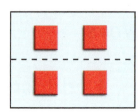 $\begin{array}{r}1\\+2\\\hline\end{array}$ 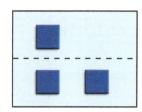 $\begin{array}{r}4\\+0\\\hline\end{array}$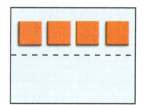

2. $\begin{array}{r}3\\+0\\\hline\end{array}$ $\begin{array}{r}3\\+1\\\hline\end{array}$ 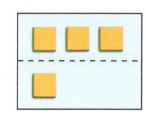 $\begin{array}{r}2\\+1\\\hline\end{array}$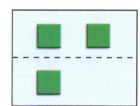

Add.

3. $1 + 3 = \underline{4}$ 4. $2 + 2 = \underline{}$

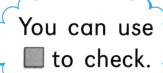

You can use ■ to check.

5. $\begin{array}{r}0\\+4\\\hline\end{array}$ $\begin{array}{r}2\\+1\\\hline\end{array}$ $\begin{array}{r}0\\+2\\\hline\end{array}$ $\begin{array}{r}1\\+0\\\hline\end{array}$ $\begin{array}{r}3\\+1\\\hline\end{array}$ $\begin{array}{r}1\\+2\\\hline\end{array}$

6. $\begin{array}{r}2\\+2\\\hline\end{array}$ $\begin{array}{r}0\\+3\\\hline\end{array}$ $\begin{array}{r}1\\+1\\\hline\end{array}$ $\begin{array}{r}0\\+1\\\hline\end{array}$ $\begin{array}{r}1\\+3\\\hline\end{array}$

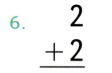

7. You buy 3 🔴 and 1 🔵.
How many collars in all?

___ collars in all

$+$ ___

66 sixty-six

Name _____

Sums of 5

Yuka tells an **addition story**.

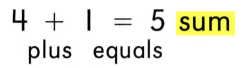

The frog sees 4 🐞.
1 more comes. There
are 5 🐞 in all.

4 + 1 = 5   **sum**
plus equals

Use 🟥 to act out each addition story.
Write the sum.

1.

   1 + 4 = 5

2.

   2 + 3 = ___

3.

   5 + 0 = ___

4.

   4 + 1 = ___

5.

   3 + 2 = ___

6.

   0 + 5 = ___

SECOND LOOK Look at the sums of 5 in 1–6.
Name the **facts** that have the same addends.

Add.

7. 2 + 3 = ___   3 + 2 = ___   1 + 4 = ___

 Ask your child to tell a story for sums of 5 such as 3+2 and 2+3.

sixty-seven **67**

Find the sum.

1. 1
   +4
   ―――
    5

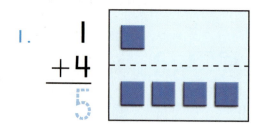

2. 2
   +3
   ―――

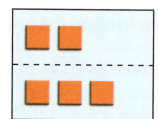

The sum is odd.

Add. Use ▪ to check.

3. 4   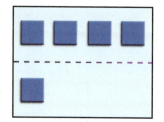   3   1   2
   +1   +2   +3   +2
   ―――   ―――   ―――   ―――
    5

4. 1   1   0   2   0
   +1   +4   +4   +1   +2
   ―――   ―――   ―――   ―――   ―――

5. 1   3   0   2   5
   +2   +1   +3   +0   +0
   ―――   ―――   ―――   ―――   ―――

6. 2   4   0   2   1
   +3   +0   +5   +2   +0
   ―――   ―――   ―――   ―――   ―――

 ✓ odd sums in 1–6.

 7. Ring the dominoes that show the sum of 5. Write the addition sentences.

___ + ___ = ___   ___ + ___ = ___

68 sixty-eight

Name _____

Sums of 6

 Both addends are the same.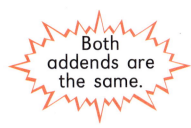

3 + 3 = 6

$\begin{array}{r} 3 \\ +3 \\ \hline 6 \end{array}$

Add.

1.

   1 + 5 = 6

2.

   5 + 1 = ___

3.

   2 + 4 = ___

4.

   4 + 2 = ___

5.

   6 + 0 = ___

6.

   0 + 6 = ___

Find the sum.

7. 4 + 2 = ___   3 + 3 = ___   2 + 4 = ___

8. 5 + 1 = ___   4 + 1 = ___   3 + 1 = ___

 9. The facts in 7 and 8 show a pattern. Describe each pattern.

Have your child tell or draw a story about 2 groups being joined to make 6 in all.

Peg has 4 🪙.
She finds 2 more.
How much does
she have now?

4¢
+2¢

6¢

She has __6__ ¢ now.

Find the sum.

1. 3¢   2¢   3¢   2¢   4¢
   +2¢   +1¢   +1¢   +4¢   +0¢
   ---   ---   ---   ---   ---
    5¢   ¢   ¢   ¢   ¢

2. 1¢   2¢   5¢   1¢   4¢
   +5¢   +2¢   +0¢   +4¢   +2¢
   ---   ---   ---   ---   ---
    ¢   ¢   ¢   ¢   ¢

3. 0   1   4   2   3
   +6   +2   +1   +3   +3
   ---   ---   ---   ---   ---

4. 1   6   1   5   0
   +3   +0   +1   +1   +3
   ---   ---   ---   ---   ---

---

5. Use 🪙. Make up a story for each fact in 1.

6. Write the missing addends. Draw pictures to check.

3 + ___ = 4 + ___

△△   △△
 △   △△

Name _____

Change the Order

Change the order.

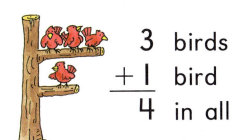   3 birds
+1 bird
———
4 in all

These are related addition facts.

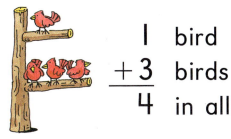   1 bird
+3 birds
———
4 in all

Add. Change the order. Draw to check.

1. 1
   +2
   ———
    3

    2
   +1
   ———
    3

2. 4
   +1
   ———

    1
   +4
   ———

3. 3
   +2
   ———

    2
   +
   ———

4. 1
   +3
   ———

    3
   +
   ———

5. 1
   +0
   ———

    +
   ———

6. 2
   +1
   ———

    +
   ———

7. 0
   +6
   ———

    +
   ———

8. 2
   +4
   ———

    +
   ———

2-6 Have your child draw a sum of 1–6 and tell you the fact. Then turn her/his drawing upside down and ask for the related addition fact.

seventy-one **71**

These are **related addition sentences**.

5 + 1 = 6

1 + 5 = 6

Color the red.

Write the related addition sentences.

1. 3 + _3_ = ___

    ___ + 3 = ___

2. 2 + ___ = ___

    ___ + 2 = ___

3. 0 + ___ = ___

    ___ + 0 = ___

4. ___ + ___ = ___

    ___ + 4 = ___

Add.

5. 4   0   0   2   4   3   5
   +2   +5   +1   +4   +1   +3   +0

6. ✔ related facts the same color in 5.

7. Write 3 facts that do not change order.

   +___   +___   +___
   ___   ___   ___
    2   4   6

72 seventy-two

DO YOU REMEMBER?

Name _____

Color each animal's path to FINISH.

 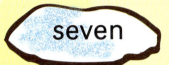  

 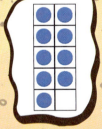 I fewer than 11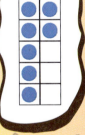

greater than 11

8, ___, 10

1 more than 8   1 more than 4   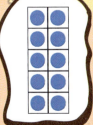   9, 8, ___

FINISH

This page reviews the mathematical content presented in Chapter 1.

seventy-three 73

Addition Patterns

Name _____

$$\begin{array}{r}1\\+3\\\hline 4\end{array} \qquad \begin{array}{r}2\\+3\\\hline 5\end{array} \qquad \begin{array}{r}3\\+3\\\hline 6\end{array}$$

Use ■ and ■ to model a pattern.

Add. Look for patterns.

1. $\begin{array}{r}1\\+0\\\hline\end{array}$ $\begin{array}{r}2\\+0\\\hline\end{array}$ $\begin{array}{r}3\\+0\\\hline\end{array}$ $\begin{array}{r}4\\+0\\\hline\end{array}$ $\begin{array}{r}5\\+0\\\hline\end{array}$ $\begin{array}{r}6\\+0\\\hline\end{array}$

2. $\begin{array}{r}5\\+1\\\hline\end{array}$ $\begin{array}{r}4\\+1\\\hline\end{array}$ $\begin{array}{r}3\\+1\\\hline\end{array}$ $\begin{array}{r}2\\+1\\\hline\end{array}$ $\begin{array}{r}1\\+1\\\hline\end{array}$

3. $\begin{array}{r}1\\+2\\\hline\end{array}$ $\begin{array}{r}2\\+2\\\hline\end{array}$ $\begin{array}{r}3\\+2\\\hline\end{array}$ $\begin{array}{r}4\\+2\\\hline\end{array}$

Add. Complete the pattern.

4. $\begin{array}{r}1\\+1\\\hline\end{array}$ $\begin{array}{r}2\\+1\\\hline\end{array}$ $\begin{array}{r}3\\+\\\hline\end{array}$ $\begin{array}{r}4\\+\\\hline\end{array}$ $\begin{array}{r}\\+\\\hline\end{array}$

5. $\begin{array}{r}4\\+2\\\hline\end{array}$ $\begin{array}{r}3\\+2\\\hline\end{array}$ $\begin{array}{r}2\\+\\\hline\end{array}$ $\begin{array}{r}\\+\\\hline\end{array}$

In 1–5 tell how the addends make a pattern. Tell how the sums make a pattern.

Name _____

Learn about Computers

Connect the dots in order. Start at 0.

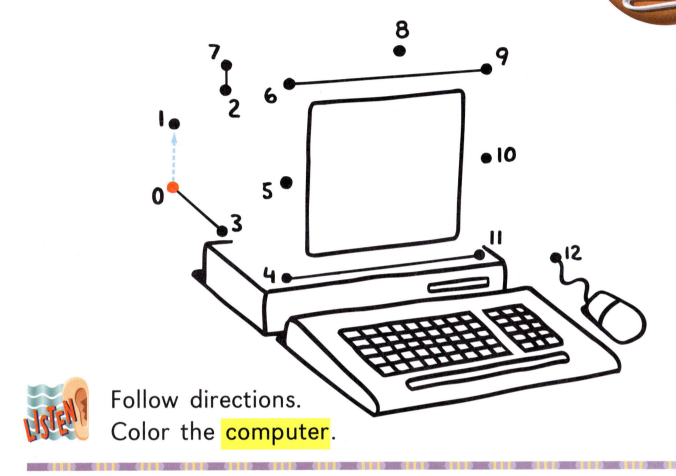

 Follow directions. Color the computer.

✔ what a computer can do.

1. count 2 4 6 8	2. walk 
3. play music 	4. drink
5. print words DOG CAT	6. draw a circle

 What else can a computer do?

You can find information on the World Wide Web.

Go to www.sadlier-oxford.com.

Click on .

 Click  Back. What happens?

Click  Forward. What happens?

 Create your own math home page.

Problem Solving   Addition Patterns

Sums of 6

76 seventy-six   Visit Sadlier on the Internet at www.sadlier-oxford.com

Name _____

PROBLEM-SOLVING STRATEGY
Use a Graph

1. Make a pictograph.
   Draw a △ for each tally.

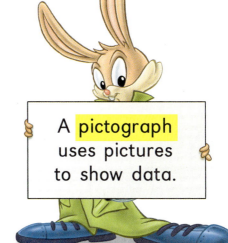

A **pictograph** uses pictures to show data.

Kind	Favorite Fun Hat
🎩(red)	△ △
🎩(black)	
🎩(yellow)	
🎩(blue)	
🎩(green)	
Key: Each △ equals 1 hat.	

Which hat is liked by most 👦?

Which hat is liked by least?

Read ➤ Draw ➤ Think ➤ Write

2. How many 👦 like 🎩 or 🎩?

   △ △ △ △ △
   △

   5
   +1
   ___

3. How many 👦 like 🎩 or 🎩?

4. How many 👦 like 🎩, 🎩, or 🎩 altogether?

Have your child make up a math story using the data from the pictograph or bar graph.

seventy-seven 77

1. Make a bar graph.
   Color a ☐ for each tally.

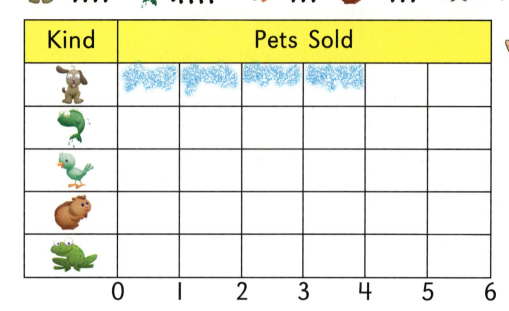

Ring your choice.

2. Did the store sell more

    or  ?

3. Were most of the pets sold

    or 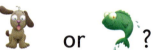 ?

4. Did the store sell as many 🐹 as 🐶 ?

   Yes or No

5. About how many pets did the store sell in all?

   more than 10
   fewer than 10

Write the number sentence to show how many in all.

6. All the and   ___ ◯ ___ = ___

7. All the 🐹 and 🐤   ___ ◯ ___ = ___

8. All the 🐸 and 🐶   ___ ◯ ___ = ___

Name _____

PROBLEM-SOLVING APPLICATIONS
Act It Out

Read → Draw → Think → Write

1. Than is first in line. Loi is last. There are 3 between them. How many are in line in all?

   __2__ ⊕ __3__ = ____

   There are ____ in all.

2. Ana has 4. She has 2 fewer than. How many does she have?

   ____ ◯ ____ = 4

   She has ____.

3. Cibi is fourth in line. 2 get in line behind her. How many are in line now?

   ____ are in line now.

4. You see 2 asleep. 2 more are playing. How many do you see altogether?

   I see ____ altogether.

5. There are 3. Then 1 more comes. How many are there in all?

   There are ____ in all.

PROBLEM SOLVING

2-10 In this lesson your child solved problems by using the *Act It Out* strategy.

seventy-nine **79**

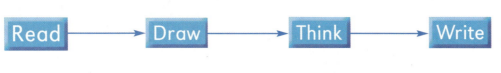

1. What hat comes next on the 🪑? Draw it.

STRATEGY FILE
Act It Out
Find a Pattern
Draw a Picture
Use a Graph

2. Jill has 3 🐟.
   She buys 3 more.
   How many fish is that in all?

   ____ fish in all

3. There are 2 gray 🐱
   and 1 white one.
   How many kittens in all?

   ____ kittens in all

4. The first 🐕 has a 🦴.
   The next 3 🐕 have 🦴.
   How many collars
   is this altogether?

   ____ collars altogether

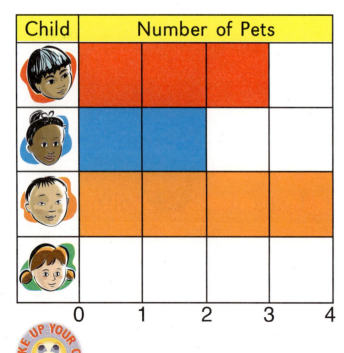

5. How many pets does the child with the most have?

   ____ pets

6. 👦 and 👧 have how many pets in all?

   ___ ◯ ___ = ___

   ____ pets

7. Write your own problem for the graph.

Name _____

Chapter Review and Practice

Find the sum.

1. $2 + 3 = \underline{5}$ $2 + 2 = \underline{}$ $2 + 1 = \underline{}$

2. $1 + 1 = \underline{}$ $1 + 5 = \underline{}$ $3 + 3 = \underline{}$

Add. Write the related addition fact.

3. $\begin{array}{r}4\\+2\\\hline\end{array}$ $\begin{array}{r}2\\+4\\\hline\end{array}$ $\begin{array}{r}1\\+3\\\hline\end{array}$ $\begin{array}{r}3\\+2\\\hline\end{array}$ $\begin{array}{r}3\\+0\\\hline\end{array}$ $\begin{array}{r}1\\+4\\\hline\end{array}$

4. $\begin{array}{r}5\\+1\\\hline\end{array}$ $\begin{array}{r}4\\+0\\\hline\end{array}$ $\begin{array}{r}2\\+4\\\hline\end{array}$ $\begin{array}{r}1\\+2\\\hline\end{array}$ $\begin{array}{r}0\\+6\\\hline\end{array}$

Color to show different sums of 5.
Write each addition sentence.

5. 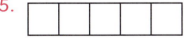 ___ + ___ = ___

   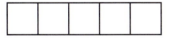 ___ + ___ = ___

   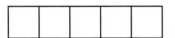 ___ + ___ = ___

PROBLEM SOLVING Draw or act out each.

6. There are 2 🐸 in a pond.
   Then 2 more jump in.
   How many 🐸 now?

   ___ ○ ___ = ___

7. I am first in line.
   There are 4 🧍 behind me.
   How many 🧍 in line in all?

   ___ ○ ___ = ___

Name _____

You buy erasers at the school store.
How much do you spend?

 1¢   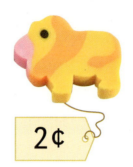 2¢    3¢    4¢

Ring the 🪙 you need.
Write the addition sentence.

1. You buy 🐞 and 🦁.

__1__¢ + __2__¢ = ____¢

2. You buy 🦁 and 🐧.

____¢ + ____¢ = ____¢

3. You buy 🦁 and 🐟.

____¢ + ____¢ = ____¢

4. You buy 🐟 and 🐟.

____¢ + ____¢ = ____¢

How many different ways can you spend each amount?

5. Exactly 4¢.

6. Less than 4¢.

Name _____

Performance Assessment

Make a bar graph to show how many letters are in each name.

TIP   KING   BO   MAX   LUCKY

1.
Name	Number of Letters					
TIP						
KING						
BO						
MAX						
LUCKY						

0   1   2   3   4   5   6

2. Each letter costs 1 . You have 6 . Which two names can you make?

Add. Does 3 or 4 make a pattern?

3. 5   3   1
   +1   +1   +1
   ──   ──   ──

4. 2   1   2
   +0   +4   +2
   ──   ──   ──

 Choose 1 of these projects. Use a separate sheet of paper.

5. Use ☐ to make dominoes. Show all the sums of 5. Write the addition sentences.

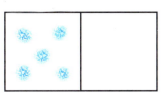 5 + 0 = 5

6. Use each 🤠 once. Make 2 addition sentences.

This page provides a variety of informal assessment opportunities in order to measure your child's understanding of Chapter 2.

eighty-three   83

Check Your Mastery

Name _____

Write the addition sentence.

1.

___ + ___ = ___

2.

___ + ___ = ___

3.

___ + ___ = ___

4.

___ + ___ = ___

Add. Write the related addition fact.

5. 5   3   2   2
   +1   +2   +4   +1
   ──   ──   ──   ──

6. 0   3   5   4
   +6   +0   +0   +0
   ──   ──   ──   ──

Write sums of 4.

7. 2   _   _   _   _
   +2   +_   +_   +_   +_
   ──   ──   ──   ──   ──
    4   4   4   4   4

PROBLEM SOLVING   Use a strategy you have learned.

8. There are 3 🐦 and 3 🐥 in a cage. How many birds in all?

___ ◯ ___ = ___

9. I am second in line. 3 more 🧍 come. How many 🧍 are in line?

___ ◯ ___ = ___

84 eighty-four

Understanding Subtraction Facts to 6

3

CRITICAL THINKING

On the way home, Kate counted 8 🔵 on bicycles. How many bicycles did Kate see?

Math Alive at Home

Dear Family,

Today your child began Chapter 3. As she/he studies subtraction facts to 6, you may want to reread the poem below, which was read in class, to her or him. Have your child talk about some of the math ideas pictured on page 85.

Look for the 🏠 at the bottom of each skills lesson. The suggestion on the page gives you an opportunity to improve your child's understanding of math and to reinforce her/his math language. You may want to have pennies and small countables available for your child to use throughout this chapter.

For more information about Chapter 3, visit the Family Information Center at www.sadlier-oxford.com

Home Activity

Pasta Parts

Try this activity with your child. You need a brown paper bag and a box of macaroni or other pasta. As your child learns subtraction facts to 6, have her/him decorate one bag for the subtraction facts for each given number, 4 through 6. After your child finishes each bag, have her/him tell you each subtraction fact she/he illustrated.

Home Reading Connection

At the Table

Milk and cookies after school
make homework fun to do.

I dunk
subtract
and take a bite

and carry over two.

Constance Andrea Keremes

Name _____

MATH CONNECTIONS
Visual Reasoning

Color the stickers that are just the **same**.

Which one is **different**? X it.

What is next in each **pattern**? Draw it.

You can put this in your Math Portfolio.

eighty-seven

CROSS-CURRICULAR CONNECTIONS
Social Studies

Name _____

What does each school helper use? Match.

Use clues. Paste the school helpers in order.

1. The  is not in the middle.
   The  is first.

2. The  is not last.
   The  is behind the .

first   middle   last

first   middle   last

88 eighty-eight

You can put this in your Math Portfolio.

Understanding Subtraction

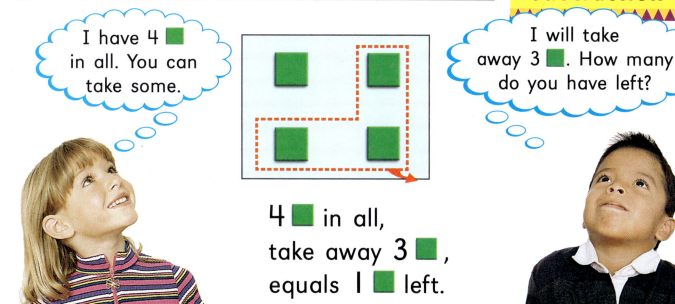

4 ◼ in all,
take away 3 ◼,
equals 1 ◼ left.

 Count to find how many ☐ in all.
Color the ☐ taken away 🖍 red.
Write how many ☐ left.

1.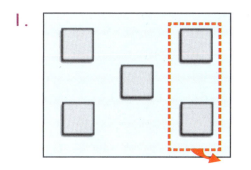

   __5__ ☐ in all,

   take away __2__ ◼,

   equals ____ ☐ left.

2.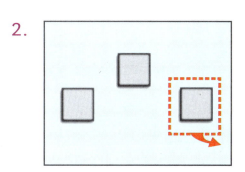

   ____ ☐ in all,

   take away ____ ◼,

   equals ____ ☐ left.

3.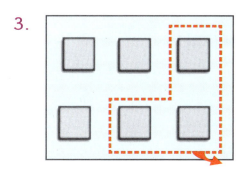

   ____ ☐ in all,

   take away ____ ◼,

   equals ____ ☐ left.

 Put a total of 6 or fewer countable objects into a clear plastic bag. Take away some of them. Ask your child to tell you the subtraction story.

eighty-nine

Use 🪙. Write the subtraction story.

1.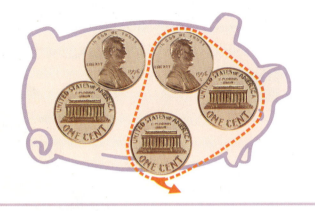

   In all __5__ ¢.

   Take away __3__ ¢ equals

   ____ ¢ left.

2.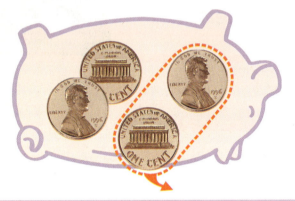

   In all ____ ¢.

   Take away ____ ¢ equals

   ____ ¢ left.

3.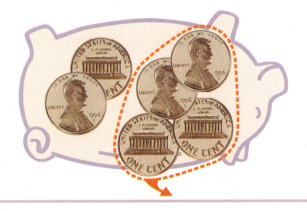

   In all ____ ¢.

   Take away ____ ¢ equals

   ____ ¢ left.

4.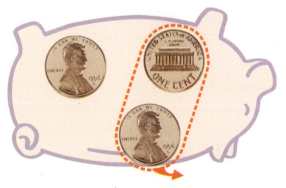

   In all ____ ¢.

   Take away ____ ¢ equals

   ____ ¢ left.

 5. Take away all. Draw and write the subtraction story.

Name _____

Subtraction Sentences

4 − 1 = 3 is a **number sentence**.

4 −̲ 1 =̲ 3
  minus **equals**

Subtract.

1.

   2 − 1 = __1__

2.

   4 − 1 = ___

3.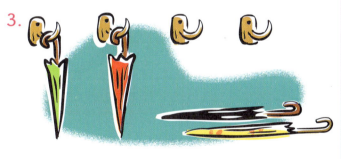

   4 − 2 = ___

4.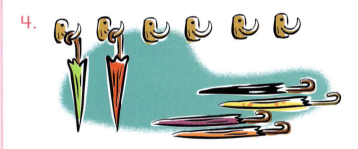

   6 − 4 = ___

5.

   3 − 2 = ___

6.

   5 − 1 = ___

 3-2 Have your child use 5 pennies to both tell a subtraction story and write a number sentence using the symbols − and =.

ninety-one **91**

Subtract. Fill in each subtraction sentence.

1.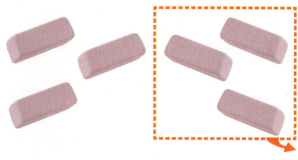

   6 − 3 = 3

2.

   ___ − ___ = ___

3.

   ___ − ___ = ___

4.

   ___ − ___ = ___

5.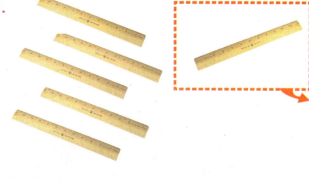

   ___ − ___ = ___

6.

   ___ − ___ = ___

7.

   ___ − ___ = ___

8.

   ___ − ___ = ___

Name _____

Subtract from 5

Carl tells a **subtraction story**.

There are 5 🚗.
1 goes away.
There are 4 🚗 left.

5 − 1 = 4   **difference**
minus equals

Use ▪ to act out each subtraction story.
Write the difference.

1.

    5 − 3 = __2__

2.

    5 − 2 = ___

3.

    5 − 5 = ___

4.

    5 − 0 = ___

5.

    5 − 1 = ___

6.

    5 − 4 = ___

Subtract.

7. 5 − 2 = ___   4 − 2 = ___   5 − 3 = ___

 In 1 and 2 how do the number taken away and the number left change?

Have your child tell a subtraction story for 5 − 5 and 5 − 0.

ninety-five **95**

Find the difference.

1. 5
  −4
  ___
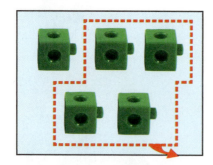

2. 5
  −1
  ___
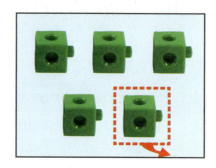

Subtract. You can use ⬛ to check.

3. 3   5   4   5
  −1   −5   −1   −3
  ___   ___   ___   ___

4. 2   4   5   3   4
  −0   −2   −2   −2   −3
  ___   ___   ___   ___   ___

5. 4   5   2   4   5
  −4   −4   −1   −0   −0
  ___   ___   ___   ___   ___

6. 1   3   5   3   2
  −0   −3   −2   −0   −2
  ___   ___   ___   ___   ___

 Ring to take away. Subtract.

7. 5 − 5 = ___

8. 5 − 3 = ___

96 ninety-six

Name

Subtract from 6

Whole − Part = Part

6 − 3 = 3

$$\begin{array}{r}6\\-3\\\hline 3\end{array}$$

Subtract.

1.
6 − 5 = ___

2. 6 − 1 = ___

3. 6 − 2 = ___

4. 6 − 4 = ___

5.
6 − 6 = ___

6.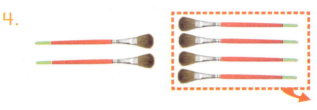
6 − 0 = ___

Find the difference.

7. 6 − 3 = ___   6 − 1 = ___   6 − 4 = ___

SHARE YOUR THINKING In 7 what numbers stand for the parts taken away? What numbers stand for the parts left?

Start with 6 pennies. Take away 6. Ask your child to say the subtraction sentence. Repeat by taking away 5, 4, 3, and so on.

ninety-seven **97**

6¢
−4¢
―――
2¢

Bill has ____ ¢ left.

Find the difference.

1. 6¢   6¢   4¢   5¢   5¢   6¢
   −3¢   −0¢   −2¢   −4¢   −5¢   −2¢
   ―――   ―――   ―――   ―――   ―――   ―――
    3 ¢   ¢   ¢   ¢   ¢   ¢

2. 5   6   4   3   5   6
   −3   −1   −3   −2   −1   −4
   ―――   ―――   ―――   ―――   ―――   ―――

3. 4   5   6   3   4   6
   −1   −2   −6   −1   −0   −5
   ―――   ―――   ―――   ―――   ―――   ―――

 Act it out. Ring Yes or No.

4. You had 6 🪙.
   Now you only have 3 🪙.
   Did you lose 3 🪙?

   Yes   No

5. I had 5 🪙.
   Now I have only 1 🪙.
   Did I lose 3 🪙?

   Yes   No

Name _____

1. Draw more to make 6 of each in all.
   Write the facts.

 3 + ___ = ___

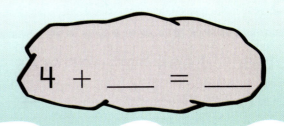 4 + ___ = ___

2. Draw more to make 5 of each in all.
   Write the facts.

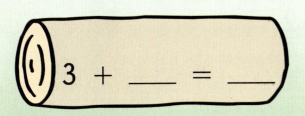 3 + ___ = ___

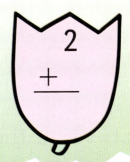 2 + ___

3. Draw more to make 4 of each in all.
   Write the facts.

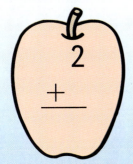 2 + ___

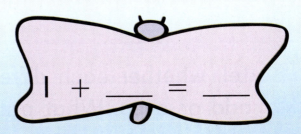 1 + ___ = ___

ninety-nine 99

Subtraction Patterns

Name _____

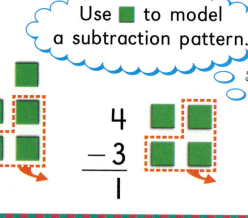

Use ■ to model a subtraction pattern.

```
  6        5        4
 -3       -3       -3
 ---      ---      ---
  3        2        1
```

Subtract. Look for patterns.

1. 6   5   4   3   2   1
   -0   -0   -0   -0   -0   -0
   --- --- --- --- --- ---
    6

2. 2   3   4   5   6
   -1   -1   -1   -1   -1
   --- --- --- --- ---

3. 6   5   4   3
   -2   -2   -2   -2
   --- --- --- ---

Subtract. Write the next fact.

4. 6   5   4   3   __
   -1   -1   -_   -_   -__
   --- --- --- --- ---

5. 3   4   5   __
   -2   -2   -_   -__
   --- --- --- ---

 Tell whether each difference in 1 and 2 is odd or even. What pattern do you see?

100 one hundred

Have your child tell how each of the patterns above was made. Then have him/her make up a pattern.

3-6

Name _____

Relate Addition and Subtraction

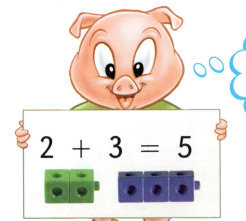

These related facts are **related number sentences**.

$2 + 3 = 5$

$5 - 3 = 2$

 What do you notice about the part joined and the part taken away?

Write the related number sentences.

1. $2 + 2 = \underline{4}$

   $4 - \underline{2} = \underline{2}$

2. $1 + 5 = \underline{6}$

   $6 - \underline{} = \underline{}$

3. $4 + 2 = \underline{}$

   $6 - \underline{} = \underline{}$

4. $3 + 3 = \underline{}$

   $6 - \underline{} = \underline{}$

5. $1 + 3 = \underline{}$

   $4 - \underline{} = \underline{}$

6. $0 + 4 = \underline{}$

   $4 - \underline{} = \underline{}$

Subtract. What is the related addition fact?

7. $\begin{array}{r}4\\-1\\\hline 3\end{array}$ $\begin{array}{r}3\\+1\\\hline\end{array}$ $\begin{array}{r}6\\-0\\\hline\end{array}$ $\begin{array}{r}3\\-1\\\hline\end{array}$ $\begin{array}{r}5\\-4\\\hline\end{array}$

3-7 Have your child tell how addition and subtraction are related by modeling $3 - 1$ and $2 + 1$.

one hundred one **101**

Zero in Facts

Name _____

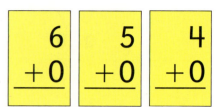

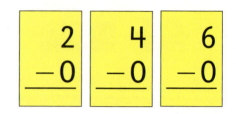

Write the number sentence. Use ▢.

1. 5 ▪ in all. Take away 0. How many ▪ left?

   5 ◯ 0 = ___

2. 5 ▪ in all. Take away all. How many left?

   ___ ◯ ___ = ___

3. 6 ▪ in all. Take away 0. How many ▪ left?

   ___ ◯ ___ = ___

4. Show 4 ▪ and 0 ▪. How many ▪ in all?

   ___ ◯ ___ = ___

Add or subtract.

5. 4   3   6   1   0   5
   −4   −0   −6   −0   +3   +0

FINDING TOGETHER

How many different number sentences can you write for each?

6.  − 0 =    − = 0   + 0 =

102 one hundred two

Give your child 6 or fewer countables. Ask your child to model a subtraction story for taking away all and then another taking away 0.

3-8

Name _____

PROBLEM-SOLVING STRATEGY
Choose the Operation

Read → Model → Ring → Write

1. I have 4 ✏️.
   I lose 3.
   How many ✏️ do I have now?

   sum   (difference)

   _4_ ⊝ _3_ = ___

   I have ___ ✏️.

2. Liam has 3 ⭐.
   He gets 2 more.
   How many ⭐ does he have now?

   (sum)   difference

   _3_ ⊕ ___ = ___

   Liam has ___ ⭐ now.

3. Terri makes 2 🥪.
   She makes 2 more.
   How many 🥪 in all?

   sum   difference

   ___ ◯ ___ = ___

   ___ 🥪 in all.

4. You see 5 🚒.
   4 go away.
   How many 🚒 are left?

   sum   difference

   ___ ◯ ___ = ___

   ___ 🚒 is left.

5. Feng has 4 🍎.
   He gives 1 away.
   How many 🍎 are left?

   sum   difference

   ___ ◯ ___ = ___

   ___ 🍎 are left.

6. 6 🧍 are in the 🚌.
   2 get off.
   How many 🧍 are left?

   sum   difference

   ___ ◯ ___ = ___

   ___ 🧍 are left.

PROBLEM SOLVING

3-9 Have your child use counters to model problems like these for addition and subtraction facts up to 6.

one hundred three **103**

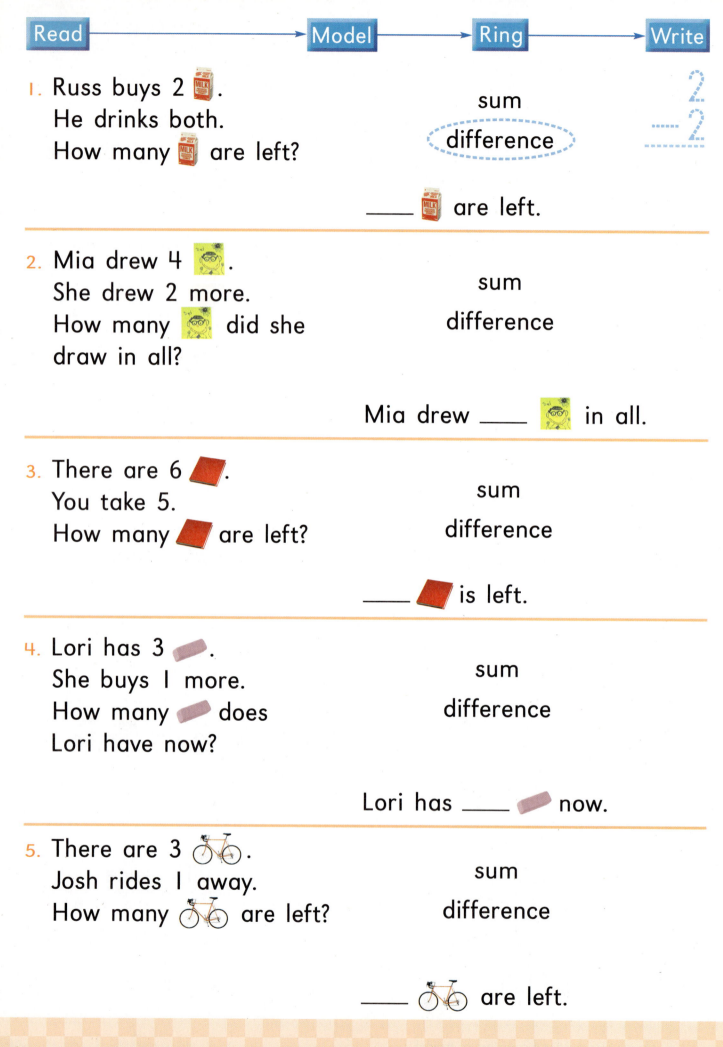

Name _____

PROBLEM-SOLVING APPLICATIONS
Write a Number Sentence

Read → Draw → Think → Write

1. There are 4 🏃 on swings. 1 more joins them. How many 🏃 in all?

   $\underline{4} \oplus \underline{1} = \underline{}$

   There are ____ 🏃 in all.

2. You have 6 ✏. You lose 3. How many ✏ do you have now?

   $\underline{6} \ominus \underline{3} = \underline{}$

   There are ____ ✏ now.

3. There are 5 🏃 on the bus. Then 1 more gets on. How many 🏃 in all?

   ___ ○ ___ = ___

   There are ____ 🏃 in all.

4. There are 5 ✶. You pick up 3. How many ✶ are left to pick up?

   ___ ○ ___ = ___

   ____ ✶ are left to pick up.

5. There are 4 🚌 in a line. The first 2 leave. How many 🚌 are left?

   ___ ○ ___ = ___

   There are ____ 🚌 left.

PROBLEM SOLVING

3-10 In this lesson your child solved problems by writing a number sentence.

one hundred five **105**

Use a strategy you have learned.

Ring to subtract.
Fill in the missing numbers.

STRATEGY FILE
Act It Out
Draw a Picture
Choose the Operation
Write a Number Sentence

PROBLEM SOLVING

1.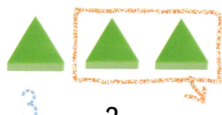

   3 − 2 = ___

   ___ ▲ is left.

2.

   ___ − 6 = ___

   ___ ⬡ are left.

3.

   ___ − 0 = ___

   ___ ◼ are left.

4.

   ___ − 4 = ___

   ___ ◆ are left.

5. Emma has 4 🍎.
   She buys 2 more.
   How many 🍎 in all?

   There are ___ 🍎 in all.

6. There are 5 🚌.
   3 pull away.
   How many 🚌 are left?

   There are ___ 🚌 left.

 Draw ___ ▲.

Take away ___ ▲.

How many ▲ are left?

___ ▲ are left.

106 one hundred six

Name _____

Chapter Review and Practice

Subtract. Use ■ to check.

1. 6 − 1 = 5   4 − 4 = ___   4 − 2 = ___
2. 5 − 3 = ___   3 − 1 = ___   4 − 0 = ___
3. 6 − 6 = ___   5 − 2 = ___   4 − 3 = ___
4. 6 − 0 = ___   2 − 0 = ___   3 − 3 = ___

Find the difference.
Write the related addition fact.

5.   6   2   4   6   3
    −4   +4   −1   −5   −2
    ──   ──   ──   ──   ──
     2   6

6.   5   5   6   5
    −1   −0   −2   −5
    ──   ──   ──   ──

7.   3   2   6   5
    −0   −1   −3   −4
    ──   ──   ──   ──

PROBLEM SOLVING

Write the number in all.
Ring to subtract.
Write the missing numbers.

8.

___ − 3 = ___

___ are left.

9.

___ − 4 = ___

___ 🟩 is left.

Name _____

Performance Assessment

Model. Write the next fact.

1. 2 – 2 = ___
   3 – 3 = ___
   4 – 4 = ___
   ___ ◯ ___ = ___

2. 6 – 0 = ___
   6 – 1 = ___
   6 – 2 = ___
   ___ ◯ ___ = ___

Ring the greater difference. What do you see?

3. 6 – 4 = ___
   6 – 2 = ___

4. 5 – 1 = ___
   5 – 4 = ___

5. 4¢   4¢
   –3¢ –1¢
   ___¢ ___¢

PORTFOLIO Choose 1 of these projects. Use a separate sheet of paper.

6. uses a ←┼┼┼→ to show 5 – 1 = 4.

 uses a ←┼┼┼→ to show 4 + 1 = 5.

Which ←┼┼┼→ did each make? Match.

   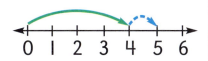

   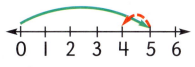

Make your own ←┼┼┼→ to add 1 and subtract 1.

7.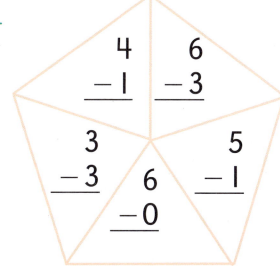

Color the difference:
- between 2 and 4
- less than 1
- just before 5

Make up your own.

Check Your Mastery

Name _____

Subtract.

1. 5 − 4 = ___   2 − 2 = ___   6 − 3 = ___

2. 6 − 6 = ___   3 − 0 = ___   5 − 2 = ___

3. 4 − 0 = ___   6 − 1 = ___   3 − 2 = ___

Find the difference. Write the related addition fact.

4.
```
  6      5      3      6      5
 −5     −3     −3     −2     −0
```

5.
```
  6      4      2      5      6
 −0     −3     −1     −5     −4
```

6.
```
  5      1      4      3      4
 −1     −0     −2     −1     −1
```

7.
```
  4      6      1      4      2
 −4     −3     −1     −0     −0
```

PROBLEM SOLVING

8. You have 2 .
You buy 2 more.
How many in all?

You have ___ in all.

9. You have 4 .
You spend all of them.
How many do you have left?

You have ___ left.

Name _____

Cumulative Review I
Chapters 1–3

Mark the ◯ for your answer.

Listening Section

A

I fewer   I more   as many as
◯   ◯   ◯

B

◯   ◯   ◯

---

1. What number comes next?

   12, 11, 10, ___

   4   5   9   7
   ◯   ◯   ◯   ◯

2. What number comes just before?

   ___, 8

   3   5   7   9
   ◯   ◯   ◯   ◯

3. What number comes between?

   0, ___, 2

   1   3   4   5
   ◯   ◯   ◯   ◯

4. What number is missing?

   5 6 7 [?] 9 10 11 12

   5   6   7   8
   ◯   ◯   ◯   ◯

5. Which is colored?

   first   third   sixth   fifth
   ◯   ◯   ◯   ◯

6. How many in all?

   2 and 3

   4   5   6   2
   ◯   ◯   ◯   ◯

7. What is 1 fewer?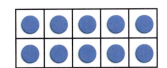

   eight   nine   ten   eleven
   ◯   ◯   ◯   ◯

8. What comes next?

   ◯   ◯   ◯   ◯

REINFORCEMENT

one hundred eleven   **111**

Mark the ○ for your answer.

9. How much altogether?

 and

1¢ 2¢ 3¢ 4¢
○   ○   ○   ○

10. Find the sum.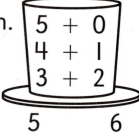

2   3   5   6
○   ○   ○   ○

11. How many are left?

1   2   3   4
○   ○   ○   ○

12. What fact comes next?
6 − 6, 5 − 5, 4 − 4, ?

3 − 3   2 − 2   3 − 0   1 − 0
  ○   ○   ○   ○

13. There are 3 🎈. All of them break. How many are left?

○ 3 − 3 = 0   ○ 3 + 3 = 6
○ 3 − 1 = 2   ○ 3 − 2 = 1

14. Dina has 2 🪙. Joe has 6 🪙. Joe has ? more than Dina.

5¢   4¢   3¢   2¢
○   ○   ○   ○

15. What fact comes next?
4 + 2, 3 + 2, 2 + 2, ?

  1   2   2   1
 +1   +0   +3   +2
 ──   ──   ──   ──
  ○   ○   ○   ○

16. Find the related addition fact for 5 − 4.

  6   5   1   5
 −4   +1   +4   −1
 ──   ──   ──   ──
  2   6   5   4
  ○   ○   ○   ○

17. How many hats in all?

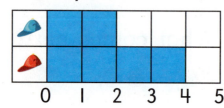

○ 2 + 4 = 6   ○ 2 + 2 = 4
○ 4 − 2 = 2   ○ 3 + 1 = 4

18. 5 🐦 in all. 2 of them fly away. How many are left?

3   4   6   7
○   ○   ○   ○

Math Alive at Home

For more information about Chapter 4, visit the Family Information Center at **www.sadlier-oxford.com**

Dear Family,

Today your child began Chapter 4. As he/she studies addition facts to 12, you may want to read the poem below, which was read in class, to him or her. Encourage your child to talk about some of the math ideas shown on page 113.

Look for the 🏠 at the bottom of each skills lesson. The suggestion on the page gives you an opportunity to improve your child's understanding of math and to reinforce his/her math language. You may want to have pennies and/or small countables available for your child to use throughout this chapter.

Home Activity

Wagon Sums

Try this activity with your child. Draw an outline of two wagons (see diagram). Place a total of 7 counters within the two wagons. Ask your child to tell an addition sentence. After your child finishes each lesson in this chapter, adjust the total number of counters and the number on the flag to correspond with sums 7 through 12.

Home Reading Connection

Rides

I ride on a bus.
I ride on a train.
I ride on a trolley.
I ride on a plane.

I ride on a ferry.
I ride in a car.
I ride on my skates—
But not very far.

But, best of all,
The ride I like
Is 'round the block
On my new bike.

Ilo Orleans

GLOSSARY

face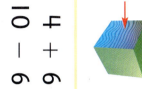

fact family
$4 + 6 = 10 \quad 6 + 4 = 10$
$10 - 6 = 4 \quad 10 - 4 = 6$

fewer

foot

fraction
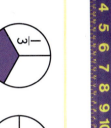
halves thirds fourths

greater than
18 14

half hour
3:30
three thirty

hour
8:00
8 o'clock

fold

inch

inside

kilogram

less than
23 36

liter

minus
— minus sign

more

nickel
5¢
or

number line
0 1 2 3 4 5

cup	(picture of cup)
cylinder	(picture of cylinder)
difference	7 − 4 = ③
digit	0, 1, 2, 3, 4, 5, 6, 7, 8, 9
dime	10¢
doubles	2 + 2
equal groups	
equal parts	
equals	= equals sign
even numbers	0, 2, 4, 6, 8, 10...

number sentence	2 + 3 = 5 addition sentence 5 − 3 = 2 subtraction sentence
odd numbers	1, 3, 5, 7, 9, 11...
on	
ordinal numbers	first, second, third
outside	
pattern	1, 2, 1, 2, 1, 2, ...
penny	1¢
pictograph	Favorite Color Butterfly

Name _____

Sums of 8

Keeshah has 4 .
She sees 4 more.
How many  in all?

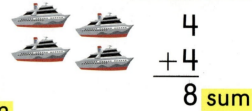

$$4 + 4 = 8 \text{ sum}$$

$$\begin{array}{r} 4 \\ +4 \\ \hline 8 \text{ sum} \end{array}$$

Find the sum.

1.

   $0 + 8 = \underline{8}$

2.

   $8 + 0 = \underline{}$

3.

   $2 + 6 = \underline{}$

4.

   $6 + 2 = \underline{}$

5.

   $7 + 1 = \underline{}$

6.

   $1 + 7 = \underline{}$

7. First complete the table.
   Then circle related facts the same color.

8 =	3	2	5	7	6	1	4
	+5	+	+	+	+	+	+

Draw an outline of a domino. Have your child draw a total of 8 dots. Ask your child to write an addition sentence that tells how many in all.

one hundred nineteen 119

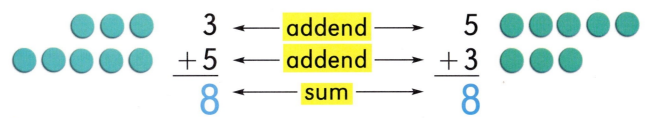

$$3 \leftarrow \text{addend} \rightarrow 5$$
$$+5 \leftarrow \text{addend} \rightarrow +3$$
$$8 \leftarrow \text{sum} \rightarrow 8$$

SHARE YOUR THINKING Are these addition facts related? How can you tell?

Add. Ring facts that belong.

	6 +1	6 +2	0 +8	4 +3	7 +1	7 +0

(6 +1 = 7 is ringed)

	1 +7	3 +3	5 +3	6 +0	4 +4	2 +5

	8 +0	5 +2	2 +4	3 +3	1 +5	1 +6

	3 +5	2 +6	4 +2	2 +3	5 +1	0 +7

PROBLEM SOLVING Use . Act out each problem.

1. Pedro has 3 .
He gets 5¢ more.
How much does he have now?

___¢ + ___¢ = ___¢

2. Jean finds 2 pennies.
She finds 6¢ more.
How much does she have in all?

___¢ + ___¢ = ___¢

120 one hundred twenty

Name _____

Sums of 9

 Listen to the addition story.

5 + 4 = 9   4 + 5 = 9

Find the sum.

1.

   2 + 7 = 9

2.

   7 + 2 = ___

3.

   3 + 6 = ___

4.

   6 + 3 = ___

5.

   0 + 9 = ___

6.

   9 + 0 = ___

CRITICAL THINKING Use ▣ or ■.

7. Which has fewer names, 9 or 7?

9 = ___ + ___ = ___ + ___ = ___ + ___ = ?

7 = ___ + ___ = ___ + ___ = ___ + ___ = ?

 Have your child use 9 objects and retell each addition story shown above.

Use or ▪ to find the sum.

```
  1            8
 +8           +1
 ―            ―
  9            9
```

TALK IT OVER Does changing the order of the addends change the sum?

Add. Ring sums less than 9.

1. 4   7   0   2
   +4   +2   +9   +5
   ―   ―   ―   ―
   8

2. 2   6   9   5   3   1
   +7   +3   +0   +4   +5   +7
   ―   ―   ―   ―   ―   ―

3. 4   3   1   6   7   3
   +5   +4   +8   +2   +1   +6
   ―   ―   ―   ―   ―   ―

4. 5   8   0   8   2   5
   +3   +1   +7   +0   +4   +2
   ―   ―   ―   ―   ―   ―

CHALLENGE How many more will balance each? Write each addition sentence.

5.

___ + ___ = ___

6.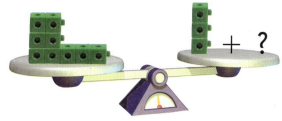

___ + ___ = ___

122 one hundred twenty-two

Name _____

Sums of 10

"I use a ▦ and ● to find a sum of 10."

5 + 5 = 10

TALK IT OVER Is the related fact for 5 + 5 = 10 the same or different?

Fill in the ▦. Add.

1.

   1 + 9 = 10

2. 9 + 1 = ___

3. 8 + 2 = ___

4. 2 + 8 = ___

5. 3 + 7 = ___

6. 7 + 3 = ___

Add. Use 🪙 and a ▦.

7. 8¢ + 2¢ = ___¢   5¢ + 4¢ = ___¢

8. 2¢ + 7¢ = ___¢   1¢ + 9¢ = ___¢

Have your child arrange 10 counters into 2 groups. Ask your child to say the addition sentence. Repeat with 7, 8, and 9.

one hundred twenty-three **123**

"The sum, 10, is even."

$$\begin{array}{r}4\\+6\\\hline 10\end{array}$$

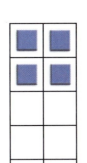 

$$\begin{array}{r}6\\+4\\\hline 10\end{array}$$

 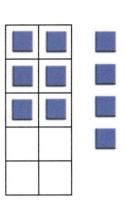

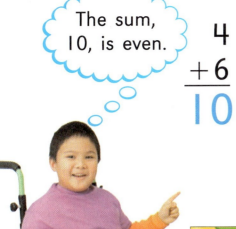

 Is each addend a part or a whole? Is the sum a part or a whole?

Find the sum.

1. $\begin{array}{r}8\\+2\\\hline 10\end{array}$ $\begin{array}{r}2\\+7\\\hline\end{array}$ $\begin{array}{r}5\\+5\\\hline\end{array}$ $\begin{array}{r}2\\+5\\\hline\end{array}$ $\begin{array}{r}7\\+3\\\hline\end{array}$ $\begin{array}{r}0\\+9\\\hline\end{array}$

2. $\begin{array}{r}7\\+2\\\hline\end{array}$ $\begin{array}{r}4\\+6\\\hline\end{array}$ $\begin{array}{r}4\\+3\\\hline\end{array}$ $\begin{array}{r}3\\+7\\\hline\end{array}$ $\begin{array}{r}5\\+4\\\hline\end{array}$ $\begin{array}{r}2\\+6\\\hline\end{array}$

3. $\begin{array}{r}6\\+0\\\hline\end{array}$ $\begin{array}{r}6\\+4\\\hline\end{array}$ $\begin{array}{r}8\\+1\\\hline\end{array}$ $\begin{array}{r}3\\+6\\\hline\end{array}$ $\begin{array}{r}2\\+8\\\hline\end{array}$ $\begin{array}{r}1\\+9\\\hline\end{array}$

4. $\begin{array}{r}6\\+3\\\hline\end{array}$ $\begin{array}{r}9\\+1\\\hline\end{array}$ $\begin{array}{r}3\\+5\\\hline\end{array}$ $\begin{array}{r}4\\+5\\\hline\end{array}$ $\begin{array}{r}0\\+8\\\hline\end{array}$ $\begin{array}{r}3\\+7\\\hline\end{array}$

 ✓ **even** sums in 1–4.

5. How much more to have 10¢? ____¢ more

Name _____

Sums of 11

8 + 3 = 11

I know 8 + 2 = 10.
So 8 + 3 = 11.

How can Dina show the related fact?

Find the sum.

1.

   4 + 7 = 11

2.

   7 + 4 = ___

3.

   9 + 2 = ___

4.

   2 + 9 = ___

5.

   5 + 6 = ___

6.

   6 + 5 = ___

Add. Use a ▦ and ■.

7. 8 + 3 = ___   1 + 9 = ___   2 + 8 = ___

8. 9 + 2 = ___   4 + 7 = ___   6 + 5 = ___

9. Name the sums of 10 that you can use in 8.

You can use 10 to find a sum of 11.

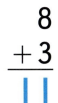
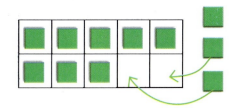

Add.

1. 7   2   5   3   8   5
  +4   +8   +6   +6   +3   +5

2. 3   2   9   1   4   6
  +4   +6   +2   +7   +5   +5

3. 3   9   2   4   3   2
  +7   +0   +9   +6   +5   +7

4. 3   2   4   1   8   6
  +8   +5   +7   +9   +2   +3

 In 1–4 ring sums of 11.

 Count on to add mentally.

5. Add 1 to: 0, 2, 4, 6, 8.
   Add 2 to: 1, 3, 5, 7, 9.
   Add 3 to: 0, 2, 4, 6, 8.

   Are these sums odd or even? _____

6. Add 1 to: 1, 3, 5, 7, 9.
   Add 2 to: 0, 2, 4, 6, 8.
   Add 3 to: 1, 3, 5, 7, 9.

   Are these sums odd or even? _____

Name _____

Sums of 12

 Tell why both children see a dozen.

Add. Color to show the related facts.

1.

   5 + 7 = ___   7 + 5 = ___

2.

   4 + 8 = ___   8 + 4 = ___

Find the sum.

3. 6 + 6 = ___   2 + 9 = ___   8 + 4 = ___

4. 5 + 7 = ___   9 + 3 = ___   7 + 5 = ___

 In 3–4 ring the related facts.

 Why does 6 + 6 not have a related fact?

Make a 6-section spinner labeled 1–6. Ask your child to spin the spinner twice and to write an addition sentence using the numbers the spinner landed on.

one hundred twenty-seven 127

Both related facts show 1 dozen in all.

```
  4              8
+ 8            + 4
───            ───
 12             12
```

Find the sum.

1. 9   5   7   5   4   2
   +3   +6   +2   +7   +4   +9
   ──   ──   ──   ──   ──   ──
   12

2. 7   6   9   6   4   3
   +4   +6   +1   +2   +3   +8
   ──   ──   ──   ──   ──   ──

3. 4   7   6   1   3   4
   +8   +3   +4   +8   +9   +7
   ──   ──   ──   ──   ──   ──

4. 6   0   5   8   8   7
   +5   +8   +4   +4   +3   +5
   ──   ──   ──   ──   ──   ──

 In 1–4 ring sums equal to 1 dozen.

 5. The 🚗 is fifth in line. There are as many cars in front of it as there are behind it. How many cars are in line?

_____ cars

Name _____

Number-Line Addition

You can use a <mark>number line</mark> to add.

$$8 + 3 = ?$$

Go to 8.   Count on 3 more.   Stop at 11.

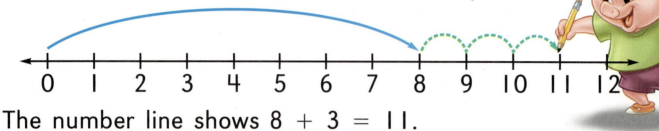

The number line shows $8 + 3 = 11$.

Show how you count on to add. Write the sum.

1. $4 + 6 =$ ___

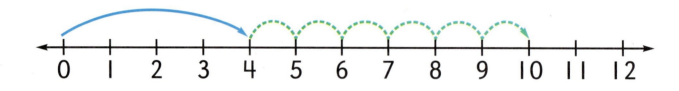

2. $9 + 2 =$ ___

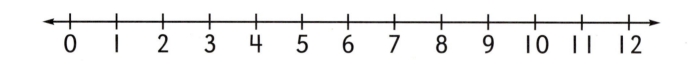

3. $6 + 5 =$ ___

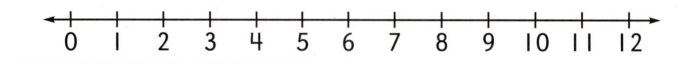

4. $5 + 7 =$ ___

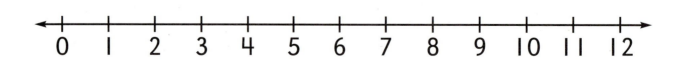

Give your child directions like the ones above. Have him/her show the action on a number line and write the addition sentence.

one hundred twenty-nine 129

Write the addition sentence for each ⬌.

1.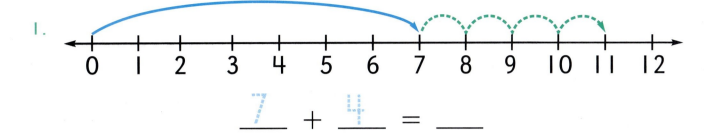

__7__ + __4__ = ___

2.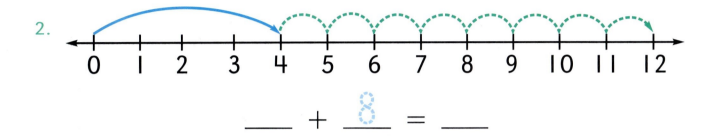

___ + __8__ = ___

3.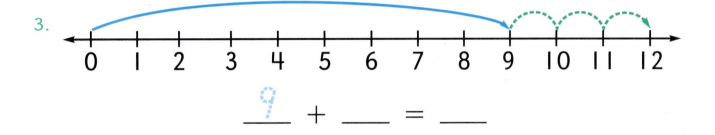

__9__ + ___ = ___

4.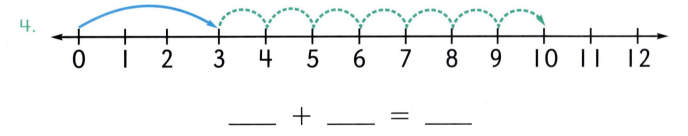

___ + ___ = ___

FINDING TOGETHER

5. Use a ⬌. Show 4 ways you can count on to add.

6. Write the number sentences.

___ + ___ = ___   ___ + ___ = ___

___ + ___ = ___   ___ + ___ = ___

Name _____

Add: Use Patterns

You can use to make an addition pattern.

5 + 6 = 11
4 + 7 = 11
3 + 8 = 11

 How does the first part joined change? How does the second part joined change? What happens to the sum?

Remember:
Part + Part = Whole

Add. Look for the pattern.
Write the next addition sentence.

1. 2 + 7 = 9
    3 + 6 = 9
    4 + 5 = 9
    5 + ___ = ___

2. 9 + 1 = ___
    8 + 2 = ___
    7 + 3 = ___
    ___ + ___ = ___

3. 3 + 6 = ___
    4 + 6 = ___
    5 + 6 = ___
    ___ + 6 = ___

4. 9 + 3 = ___
    8 + 3 = ___
    7 + 3 = ___
    ___ + ___ = ___

5. 4 + 8 = ___
    4 + 6 = ___
    4 + 4 = ___
    ___ + ___ = ___

6. 3 + 3 = ___
    3 + 5 = ___
    3 + 7 = ___
    ___ + ___ = ___

Ask your child to tell how each pattern above was made. Let her/him make up a pattern for you.

Find the sum. Complete the pattern.

1.
   9 8 7 6
   +3 +4 +5 +6 +___
   12 12

2.
   2 4 6 8
   +2 +2 +2 +2 +___

3.
   3 3 3
   +0 +2 +4 +___ +___

4.
   6 7 8
   +5 +4 +3 +___ +___

5. 5 + 2 = ___
 4 + 3 = ___
 3 + 4 = ___
 ___ + ___ = ___

6. 5 + 5 = ___
 4 + 4 = ___
 3 + 3 = ___
 ___ + ___ = ___

Look for patterns.

7. Make up a pattern for each.

 ___ + 1 = ___ ___ + 0 = ___

8. Write and draw an addition pattern that you like.

Name _____

Doubles and Near Doubles

These are doubles facts.
In a doubles fact,
both parts
are the same.

Remember:
Part + Part = Whole

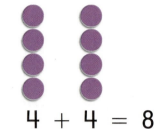

1 + 1 = 2   2 + 2 = 4   3 + 3 = 6   4 + 4 = 8

Draw the part joined in each doubles fact.
Complete the addition sentence.

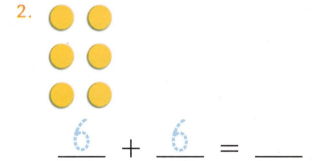

1. 5 + 5 = 10

2. 6 + 6 = ___

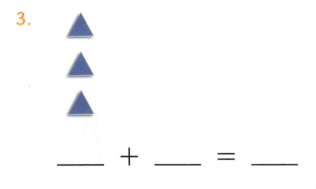 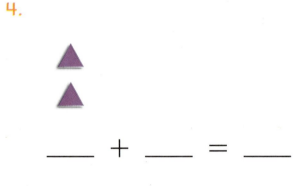

3. ___ + ___ = ___

4. ___ + ___ = ___

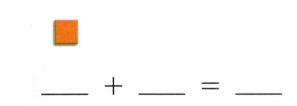 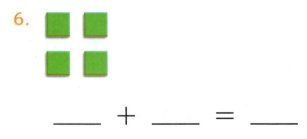

5. ___ + ___ = ___

6. ___ + ___ = ___

SHARE YOUR THINKING

You have 1 **pair** of legs. A 🐱 has 2 pairs.
How many pairs of legs does a 🕷 have?

Have your child cut out pictures of objects from a magazine to show doubles. Then have him/her write the addition sentence.

one hundred thirty-three **133**

You can use a doubles fact to add 3 + 4.

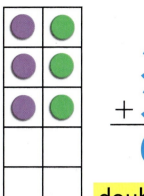
$$\begin{array}{r}3\\+3\\\hline 6\end{array}$$
doubles

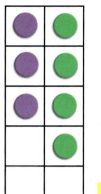
$$\begin{array}{r}3\\+4\\\hline 7\end{array}$$
doubles plus 1

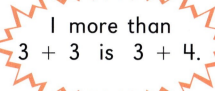
1 more than 3 + 3 is 3 + 4.

Write the doubles fact. Draw one more.
Write the new addition fact.

1. 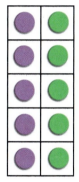 $\begin{array}{r}5\\+5\\\hline 10\end{array}$ $\begin{array}{r}5\\+6\\\hline 11\end{array}$

2. 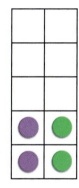 ___ + ___   ___ + ___

3. 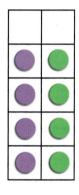 ___ + ___   ___ + ___

4. 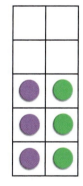 ___ + ___   ___ + ___

Add. ✔ when you used a doubles fact.

5. 7 + 5 = ___   8 + 3 = ___   6 + 6 = ___

6. 4 + 7 = ___   6 + 5 = ___   3 + 2 = ___

 CHALLENGE How can you use a doubles fact to add 6 + 7?

Name _____

Adding Three Numbers

Here are two ways to **group** addends.

You can add down and group 2 + 4 first.

Or you can add up and group 4 + 3 first.

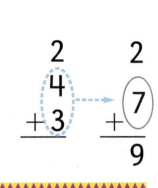

Add. You can use .

1.
(2
3)→(5)
+4

 +4
 9

 (3
 3)→( )
 +5

 +5

 (4
 5)→( )
 +3

 +3

2.
2
(2
+6)→( )

 2
+( )

 2
(7
+2)→( )

 2
+( )

 2
(5
+2)→( )

 2
+( )

3.
2
5
+1

 4
 0
+5

 3
 2
+7

 2
 5
+4

 2
 3
+2

4.
4
1
+4

 3
 3
+6

 5
 3
+0

 3
 1
+6

 4
 1
+5

4-10 Ask your child to describe how she/he finds the sum of three addends.

Group doubles first.

2 + (4 + 4) = ___   (3 + 3) + 4 = ___

2 + 8 = 10   6 + 4 = 10

 How are these addition sentences the same?

Find the sum.
Ring the numbers you add first.

1. 2 + 2 + 6 = ___   2. 3 + 4 + 4 = ___

   ___ + ___ = ___   ___ + ___ = ___

3. 3 + 2 + 6 = ___   4. 5 + 5 + 0 = ___

   ___ + ___ = ___   ___ + ___ = ___

5. 3 + 3 + 4 = ___   6. 7 + 4 + 1 = ___

   ___ + ___ = ___   ___ + ___ = ___

 Add. Use sums of 10.

7.   6   2   3   2   9
    2   5   7   1   2
  +4   +5   +1   +8   +1

 How does looking for a 10 or doubles make it easier to add?

136   one hundred thirty-six

Name _____

Chain Addition

You can add 6 + 1 + 2 on a number line.

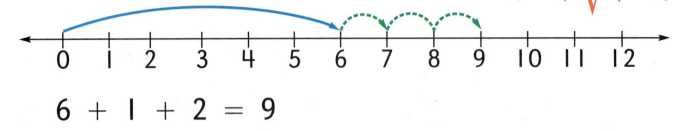

Add left to right.

Go to 6. Count on 1. Count on 2.

6 + 1 + 2 = 9

Find the sum. Use a ←—→.

1. 3 + 4 + 2 = _9_
2. 5 + 1 + 2 = ___
3. 6 + 4 + 1 = ___
4. 2 + 8 + 1 = ___
5. 3 + 3 + 2 = ___
6. 5 + 5 + 2 = ___
7. 5 + 6 + 0 = ___
8. 7 + 4 + 0 = ___

 What happens when the third addend is 0?
What happens when the third addend is 1?

Write an addition sentence for each ←—→.

9.
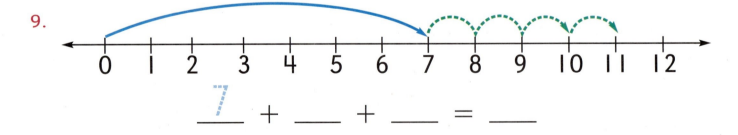

_7_ + ___ + ___ = ___

10.
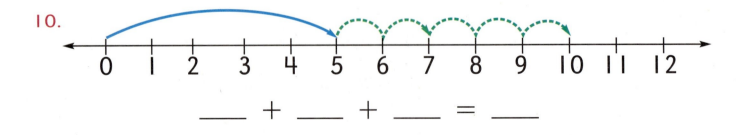

___ + ___ + ___ = ___

Have your child use a number line to show the addition sentences on page 138.

one hundred thirty-seven **137**

Write the addition sentence for each.

1. Austin has 5 .
   He buys 3 more.
   Then he buys 1 more.
   How many in all? ___ + ___ + ___ = ___

2. Sara feeds 6 . Tom feeds
   3 . Joy feeds 3 .
   How many animals fed
   in all? ___ + ___ + ___ = ___

3. Yori sells 4 , 2 ,
   and 5 . How many
   collars in all? ___ + ___ + ___ = ___

4. Seth has 5 .
   He buys 5 .
   Then he buys 1 more.
   How many fish now? ___ + ___ + ___ = ___

5. There are 6 , 1 ,
   and 5 . How many
   birds altogether? ___ + ___ + ___ = ___

 6. The sum of three addends is 12.
   One addend is 2. Find the other two.
   Tell why there is more than 1 answer.

   ```
    ☐
    ☐
   +2
   ──
   12
   ```

Name _____

Addition Strategies

You can use these addition strategies.

Count on.

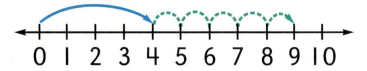

Look for patterns.

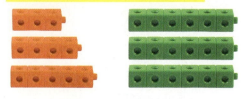

Use doubles.

Think 10.

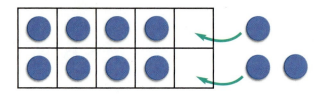

Find the sum.

1. 8   7   8   3   8¢   9¢
   +2   +4   +1   +7   +4¢   +3¢
                                ___¢   ___¢

2. 3   5   2   2   7¢   6¢
   +9   +4   +8   +9   +3¢   +4¢
                                ___¢   ___¢

3. 4   6   8   7   9¢   6¢
   +7   +6   +3   +5   +1¢   +5¢
                                ___¢   ___¢

4. 8 + 1 + 2 = ___   5¢ + 5¢ + 2¢ = ___¢

Which addition strategy did you use most?

Have your child describe how to use two of the strategies shown on this page to add.

one hundred thirty-nine **139**

Name _____

This is an **input-output** machine.

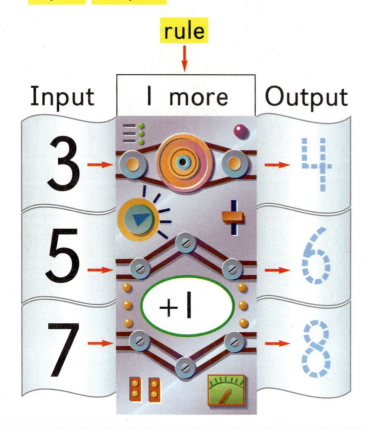

1 more than 3 is 4.

Write the missing input or output numbers.

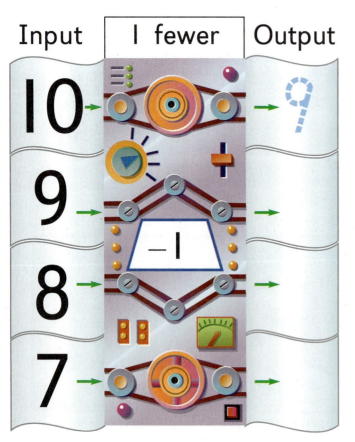
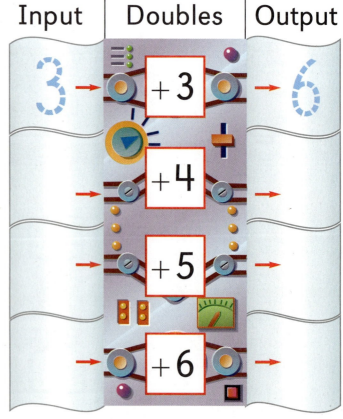

4-13 Your child learned how to apply a rule to find an output. Tell him/her to explain one of the machines shown above.

one hundred forty-one 141

Write the rule. Complete the machine.

1.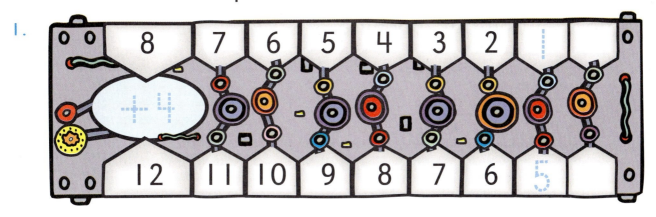

2.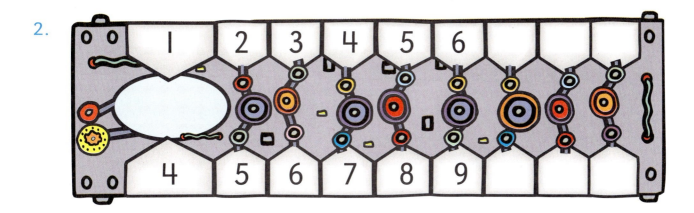

3.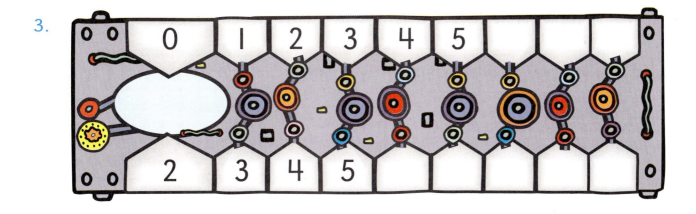

4.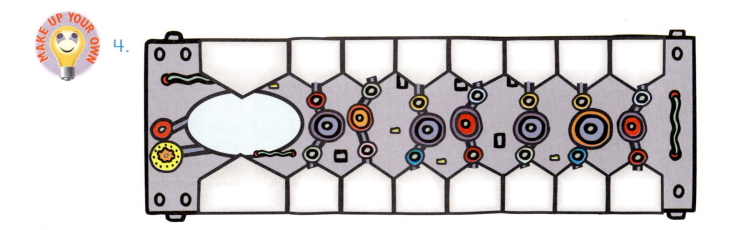

Name _____

Choose the question.
Write the number sentence.

1. **Read** — Cora sees 9 .

   Then she sees 2 more.

   Think — ● How many in all?
   ○ How many are left?

   Write — 9 + 2 = 11

   Think about joining and separating to choose the question.

2. **Read** — There are 6 at the station.

   4  leave.

   Think — ○ How many in all?
   ○ How many are left?

   Write — ___ ◯ ___ = ___

3. **Read** — There are 8 children on the .

   3 more children get on.

   Think — ○ How many children in all?
   ○ How many children are left?

   Write — ___ ◯ ___ = ___

Have your child explain how she/he knew which question to choose.

one hundred forty-three   143

ese steps. Read → Think → Write

Choose the question. Write the number sentence.

1. There are 5 ⛵.
   Then 2 sail away.

   ○ How many ⛵ in all?

   ○ How many ⛵ are left?

   ___ ○ ___ = ___

2. Dennis sees 5 🚁.
   4 more come.

   ○ How many 🚁 in all?

   ○ How many 🚁 are left?

   ___ ○ ___ = ___

3. Ruby counts 7 🚚.
   She counts 5 more.

   ○ How many 🚚 in all?

   ○ How many 🚚 are left?

   ___ ○ ___ = ___

4. I see 6 ✈.
   2 of them fly away.

   ○ How many ✈ in all?

   ○ How many ✈ are left?

   ___ ○ ___ = ___

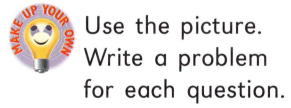 Use the picture. Write a problem for each question.

5. How many in all?

   ___ ○ ___ = ___

6. How many are left?

   ___ ○ ___ = ___

144   one hundred forty-four

PROBLEM-SOLVING APPLICATIONS
Choose the Operation

Name _____

| Read | Model | Think | Write |

1. Koko cut out 6 🚶.
   She cuts out 4 more.
   How many 🚶 did
   Koko cut out?

   (add)
   subtract

   6
   +4
   ―――
   10

   Koko cut out ____ 🚶 in all.

2. Yori has 6 ✈ to make.
   He made 3 of them.
   How many more ✈
   must Yori make?

   add
   (subtract)

   6
   −3
   ―――

   Yori must make ____ ✈.

3. The children sold 5 ⭐.
   Then they sold 6 more.
   How many ⭐ did they
   sell altogether?

   add
   subtract

   They sold ____ ⭐ altogether.

4. Luis buys 7 🚁 kits. He
   buys 2 more. How many
   🚁 kits in all?

   add
   subtract

   Luis buys ____ 🚁 kits in all.

5. Cindy makes 5 🦢.
   She gives 3 away. How
   many 🦢 are left?

   add
   subtract

   There are ____ 🦢 left.

Ask your child to tell you how he/she solved his/her favorite problem in this lesson.

one hundred forty-five 145

Use a strategy you have learned.

STRATEGY FILE
Act It Out
Ask a Question
Choose the Operation

1. The ninth 🟥 in the pattern is red. There are 3 more after it. How many cubes does the pattern have?

   ___ ◯ ___ = ___

   The pattern has ___ cubes.

2. Lisa wants to make 6 👑. She made 5. How many 👑 left to make?

   add   or   subtract

   ___ ◯ ___ = ___

   ___ 👑 left to make.

3. I cut out 8 ❤. I cut out 4 more. How many ❤ cut out in all?

   add   or   subtract

   ___ ◯ ___ = ___

   ___ ❤ cut out in all.

4. Ling has 5 🪙. He finds 5¢ more. How much does Ling have now?

   add   or   subtract

   ___¢ ◯ ___¢ = ___¢

   Ling has ___¢ now.

5. I bake ___ 🧍. I eat ___ 🧍. How many
   _____

   _____ ?

   ___ 🧍

6. I have ___ 🧍. I buy ___ more 🧍. How many
   _____

   _____ ?

   ___ 🧍

Chapter Review and Practice

Name _____

Add. Show your work on a ⟵⋯⟶.

1. $8 + 2 = \underline{10}$ $6 + 3 = \underline{}$ $5 + 5 = \underline{}$
2. $3 + 7 = \underline{}$ $1 + 7 = \underline{}$ $6 + 2 = \underline{}$
3. $3 + 2 + 4 = \underline{}$ $4 + 3 + 5 = \underline{}$
4. $2 + 2 + 7 = \underline{}$ $3 + 6 + 2 = \underline{}$

Find the sum.

5. 9   3   7   6   8   2¢
   +1   +6   +4   +6   +1   +7¢
                                 ___¢

6. 4   4   6   0   3   8¢
   +3   +4   +4   +8   +9   +4¢
                                 ___¢

7. 0   5   2   2   5   9¢
   +7   +3   +9   +6   +6   +3¢
                                 ___¢

8. 4   6   3   3   7   1¢
   +8   +5   +5   +8   +1   +9¢
                                 ___¢

PROBLEM SOLVING

9. There are 6 🛞 on a truck. 2 🛞 fall off. How many 🛞 are left? _____

10. Jay has 4 🪙. He gets 5¢. Then he finds 2¢ more. How much does Jay have now? _____¢

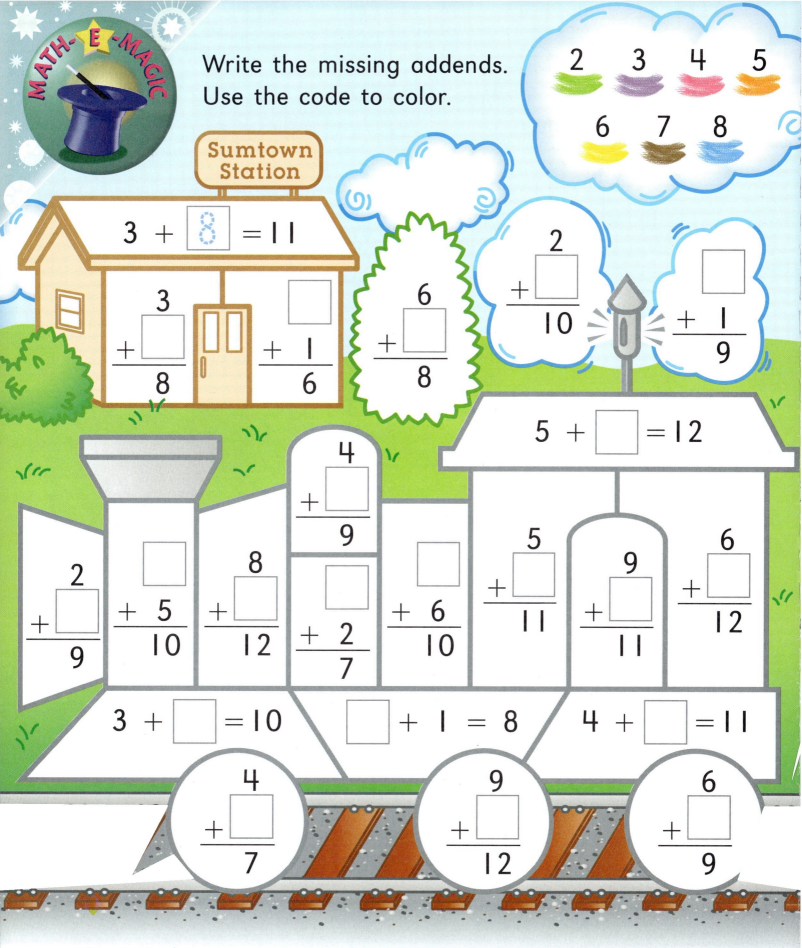

Performance Assessment

Name _____

1. Correct these sums. Use or ▇.
   Name the strategy you used.

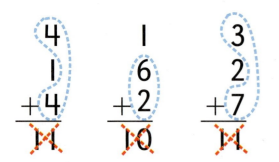

Add up or add down.
Use doubles or
sums of 10.

2. Color the greater sum in each ▭.
   How can you find it without adding?

| 5 + 5 | 5 + 6 | 5 + 4 | 4 + 4 + 1 |
| 5 + 7 | 4 + 3 | 9 + 2 | 5 + 5 + 1 |

PORTFOLIO Choose 1 of these projects. Use a separate sheet of paper.

3. Complete the addition square.
   Add → and add ↓.

   3 + □ = 6

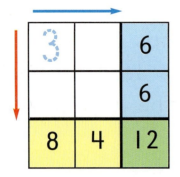

4. Spin for three addends.
   Write an addition sentence.
   Make up 3 more sentences.

   Ring the greatest sum.
   ✔ the least sum.

   1 + 1 + 2 = 4

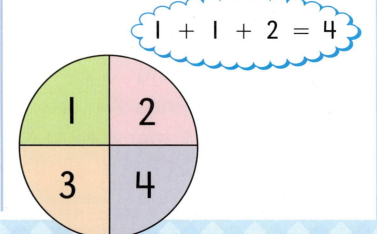

This page provides a variety of informal assessment opportunities in order to measure your child's understanding of the skills taught in Chapter 4.

one hundred forty-nine 149

Check Your Mastery

Name _____

Find the sum.

1.
```
  5      5      2      5      8      9¢
 +7     +3     +5     +6     +3    +2¢
                                    ___
                                     ¢
```

2.
```
  5      8      3      6      3     8¢
 +4     +2     +4     +2     +9    +4¢
                                    ___
                                     ¢
```

3.
```
  5      5      7      3      3     3¢
  4      0      2      4      1     3¢
 +1     +2     +3     +4     +8    +2¢
                                    ___
                                     ¢
```

Add. Use a ⇿.

4. $3 + 8 = \underline{\quad}$ $3 + 6 = \underline{\quad}$ $6 + 4 = \underline{\quad}$

Complete each pattern. Use 🟩 or 🟦.

5. $4 + 5 = \underline{\quad}$ $7 + 3 = \underline{\quad}$ $4 + 3 = \underline{\quad}$

$5 + 5 = \underline{\quad}$ $6 + 4 = \underline{\quad}$ $4 + 5 = \underline{\quad}$

$\underline{\quad} + \underline{\quad} = \underline{\quad}$ $\underline{\quad} + \underline{\quad} = \underline{\quad}$ $\underline{\quad} + \underline{\quad} = \underline{\quad}$

PROBLEM SOLVING Use a strategy you have learned.

6. Duke has 3 . He finds 7 more. How many in all?

   _____

7. Rose has 6 . She spends 3¢. How much does she have now?

   _____¢

Subtraction Facts to 12

5

CRITICAL THINKING

Alyce's favorite balloon does not have stripes. It is not round. It is not yellow. Which balloon is her favorite? Draw it.

Math Alive at Home

Dear Family,

Today your child began Chapter 5. As your child studies subtraction facts to 12, you may want to read the story below, which was read in class, with her or him. Have your child talk about some of the math ideas pictured on page 151.

Look for the 🏠 at the bottom of each skills lesson. The suggestion on the page gives you an opportunity to improve your child's understanding of math. You may want to have pennies and small countables available for your child to use throughout this chapter.

Home Activity

Make It Equal

Try this activity with your child. Put some counters on one plate. This represents the total. Fold a second paper plate in half and put fewer counters on the left side than are on the first plate. Ask your child to tell how many more counters are needed to have the same number in all on both plates. After your child finishes each lesson in this chapter, adjust the total number of counters.

For more information about Chapter 5, visit the Family Information Center at www.sadlier-oxford.com

Home Reading Connection

County Fair

Todd and Alyce had been waiting for weeks for the county fair to come to town. The day finally arrived! As the children entered the fairground, they saw 3 tents. Todd and Alyce quickly ran to the ticket booth. Todd was third in line and Alyce was in front of him. Todd bought 6 tickets. So did Alyce.

Todd and Alyce each wanted to buy a different shaped balloon from the balloon man. The children watched a little girl put her first-prize sheep, Curly, in the pen. Music from a nearby band filled the air. The children were excited by all the sights and sounds. They were glad that their family was going to spend the day together. What other things do you think happened at the fair?

Name _____

MATH CONNECTIONS
Tally and Graph

Use the picture of the county fair to fill in the chart below.

 Listen to directions.

	Tally	Number	Odd	Even
🎪				
🐑				
🚃				

Use your tally. Complete the bar graph.

Tell how many more ☐ you color.

County Fair

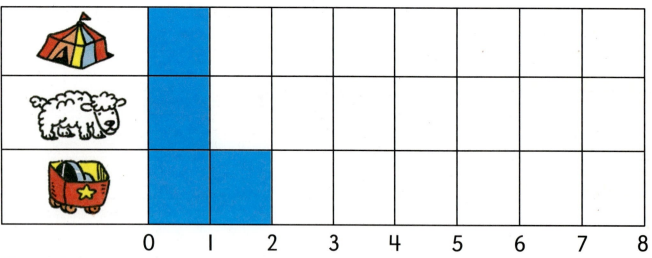

Try this!

On another sheet of paper, draw a clown.

Use these shapes: ● ▲ ■ ▬.

Tally the number of each shape you used.

 You can put this in your Math Portfolio.

one hundred fifty-three **153**

CROSS-CURRICULAR CONNECTIONS
Map Reading

Name _____

This is a map of the fairground.

1. How many ⬭ from 🎟️ to 🎯 ? _____

2. Start at 🎟️. Go to ⛺.
   How many ⬭ did you take? _____
   Describe your path.

3. Color ⬭ red for longest path from 🟫 to 🎈.

4. Draw a balloon man on the map.
   Use ⬭ to draw a path to him from 🎟️.

Name _____

Other Names: Sums and Differences

These are some names for 7.

This is a name for 7, too.

6 + 1 = 5 + 2

How would you model 7 + 0 ?

1. Write other names for 8.

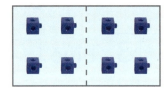  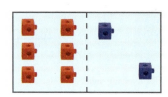 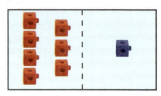

4 + 4 = ___ + ___ = ___ + ___ = ___ + ___

2. Write other names for 9.

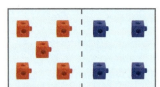 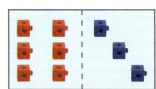 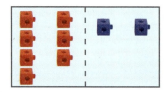 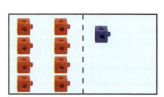

5 + 4 = ___ + ___ = ___ + ___ = ___ + ___

 3. Describe the patterns. Name the next fact for each.

 Use 🟥 or 🟦.

4. Which has more names, 10 or 11? ____

10 = ___ + ___ = ___ + ___ = ___ + ___ = ?

11 = ___ + ___ = ___ + ___ = ___ + ___ = ?

5-1 Ask your child to list and model facts for a sum of 12 and a difference of 4.

one hundred fifty-five 155

You can subtract to find other names for 3.

6 − 3 = 3
in all red blue

6 in all
−3
 3

1. Write other names for 3. Color the cubes.

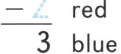 in all
   red
3 blue

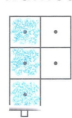

___
3 blue

___
3 blue

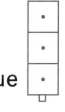

Use 🟧 or 🟦.

2. Write other names for 2.

6	5	4	3	2
−	−	−	−	−
___	___	___	___	___

3. Write other names for 1.

6	5	4	3	2
−	−	−	−	−
___	___	___	___	___

Use 🟧 or 🟦.

4. Write other names for 0. Describe the pattern.

0 = ___ − ___ = ___ − ___

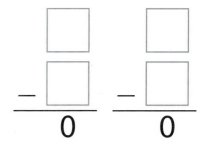

Name _____

Subtract from 7

 Listen to the story.

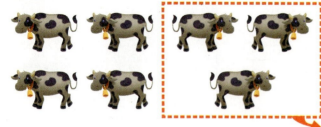 

7 − 3 = 4 difference   7 − 4 = 3 difference

Subtract. Tell how many are left.

1.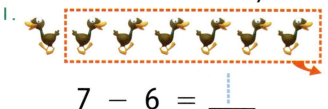

   7 − 6 = ___

2.

   7 − 1 = ___

3.

   7 − 5 = ___

4.

   7 − 2 = ___

5.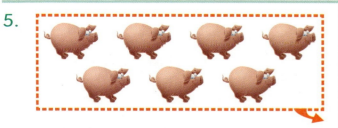

   7 − 7 = ___

6.

   7 − 0 = ___

 7 − 3 and 7 − 4 are related.
How are they alike? How are they different?

7. Write two more related subtraction sentences.

   7 − ___ = ___   7 − ___ = ___

one hundred fifty-seven **157**

Find the difference.

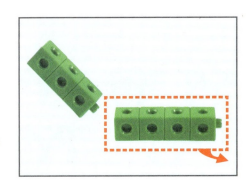
7
−4
3

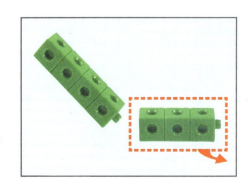
7
−3
4

These are **related subtraction facts**.

Subtract. Draw or model the related fact to check.

1. 7   6   7   5   6   7
  −1   −2   −2   −3   −5   −3
   6

2. 5   7   6   7   6   7
  −1   −4   −0   −5   −3   −2

3. 7   5   6   7   6   7
  −7   −5   −4   −6   −1   −4

4. 6   5   5   7   7
  −6   −0   −4   −5   −0

CHALLENGE Write the related subtraction sentence for each addition sentence.

5. 2 + 5 = 7   ___ − ___ = ___

6. 1 + 6 = 7   ___ − ___ = ___

158 one hundred fifty-eight

Name _____

Subtract from 8

Nicki has 8 🙂.
She gives 4 away.
How many does
she have now?

You can write
subtraction
in two ways.

8 – 4 = 4

$$\begin{array}{r}8\\-4\\\hline 4\end{array}$$

Ring the part taken away. Subtract.

1.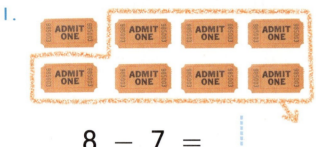

   8 – 7 = __

2.

   8 – 1 = __

3.

   8 – 5 = __

4.

   8 – 3 = __

5.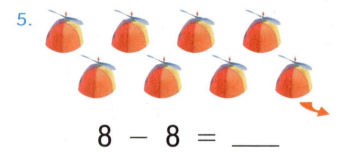

   8 – 8 = __

6.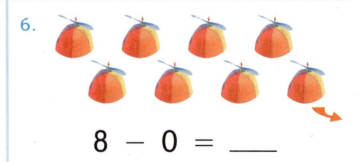

   8 – 0 = __

Find the difference.

7. 8 – 3 = __   7 – 5 = __   8 – 4 = __

 Which fact above is the same as its related subtraction sentence? Why?

 Start with 8 pennies. Take away 8. Ask your child to say the subtraction sentence. Repeat by taking away 7, 6, 5, and so on.

You can model related subtraction facts.

$$\begin{array}{r}8\\-6\\\hline 2\end{array}$$ 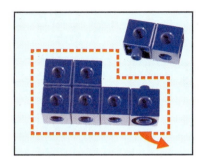 $$\begin{array}{r}8\\-2\\\hline 6\end{array}$$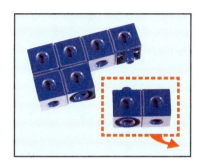

Subtract. You can use  to check.

1. $\begin{array}{r}8\\-8\\\hline 0\end{array}$ $\begin{array}{r}7\\-3\\\hline\end{array}$ $\begin{array}{r}8\\-4\\\hline\end{array}$ $\begin{array}{r}5\\-4\\\hline\end{array}$ $\begin{array}{r}8\\-6\\\hline\end{array}$

2. $\begin{array}{r}7\\-5\\\hline\end{array}$ $\begin{array}{r}8\\-3\\\hline\end{array}$ $\begin{array}{r}8\\-0\\\hline\end{array}$ $\begin{array}{r}6\\-3\\\hline\end{array}$ $\begin{array}{r}7\\-7\\\hline\end{array}$ $\begin{array}{r}8\\-1\\\hline\end{array}$

3. $\begin{array}{r}8\\-5\\\hline\end{array}$ $\begin{array}{r}6\\-6\\\hline\end{array}$ $\begin{array}{r}8\\-7\\\hline\end{array}$ $\begin{array}{r}8\\-2\\\hline\end{array}$ $\begin{array}{r}7\\-4\\\hline\end{array}$ $\begin{array}{r}7\\-0\\\hline\end{array}$

4. $\begin{array}{r}7\\-2\\\hline\end{array}$ $\begin{array}{r}6\\-5\\\hline\end{array}$ $\begin{array}{r}7\\-1\\\hline\end{array}$ $\begin{array}{r}6\\-4\\\hline\end{array}$ $\begin{array}{r}7\\-6\\\hline\end{array}$ $\begin{array}{r}6\\-2\\\hline\end{array}$

 PROBLEM SOLVING   Use 🪙. Solve.

5. You have 8 🪙. You spend 3 of them. How many 🪙 do you have now?

 now

6. I have 7 🪙. You have 8 🪙. What is the difference in our 🪙?

Name _____

Subtract from 9

Tell each subtraction story.

9 − 6 = 3 difference   9 − 3 = 6 difference

Find the difference.

1.

   9 − 8 = ___

2.

   9 − 1 = ___

3.

   9 − 5 = ___

4.

   9 − 4 = ___

5.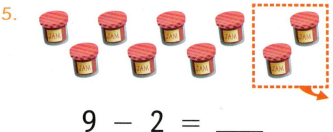

   9 − 2 = ___

6.

   9 − 7 = ___

Subtract.

7. 9 − 4 = ___   9 − 0 = ___   9 − 2 = ___

8. 9 − 6 = ___   9 − 3 = ___   9 − 9 = ___

 Ring related subtraction facts in 7–8.

Write the numbers for related facts for 9— for example, 5, 4, 9. Ask your child to model and write both facts.

one hundred sixty-one   161

You can subtract down.

$$\begin{array}{r}9\\-0\\\hline 9\end{array}$$ 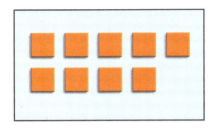 $$\begin{array}{r}9\\-9\\\hline 0\end{array}$$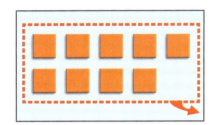

Find the difference. Ring facts that belong.

$$\begin{array}{r}8\\-3\\\hline 5\end{array}$$ $$\begin{array}{r}9\\-6\\\hline\end{array}$$ $$\begin{array}{r}8\\-4\\\hline\end{array}$$ $$\begin{array}{r}7\\-2\\\hline\end{array}$$ $$\begin{array}{r}9\\-1\\\hline\end{array}$$ $$\begin{array}{r}9\\-7\\\hline\end{array}$$

$$\begin{array}{r}7\\-6\\\hline\end{array}$$ $$\begin{array}{r}9\\-5\\\hline\end{array}$$ $$\begin{array}{r}9\\-3\\\hline\end{array}$$ $$\begin{array}{r}9\\-8\\\hline\end{array}$$ $$\begin{array}{r}8\\-2\\\hline\end{array}$$ $$\begin{array}{r}7\\-0\\\hline\end{array}$$

$$\begin{array}{r}7\\-1\\\hline\end{array}$$ $$\begin{array}{r}9\\-2\\\hline\end{array}$$ $$\begin{array}{r}8\\-5\\\hline\end{array}$$ $$\begin{array}{r}8\\-1\\\hline\end{array}$$ $$\begin{array}{r}9\\-4\\\hline\end{array}$$ $$\begin{array}{r}9\\-0\\\hline\end{array}$$

$$\begin{array}{r}8\\-8\\\hline\end{array}$$ $$\begin{array}{r}7\\-3\\\hline\end{array}$$ $$\begin{array}{r}8\\-6\\\hline\end{array}$$ $$\begin{array}{r}8\\-7\\\hline\end{array}$$ $$\begin{array}{r}9\\-9\\\hline\end{array}$$ $$\begin{array}{r}6\\-0\\\hline\end{array}$$

 PROBLEM SOLVING Color and solve.

1. Color the ninth ☐.
   How many are not colored?

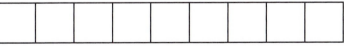

___ ◯ ___ = ___

2. Color the third and first △. How many still need to be colored?

___ ◯ ___ = ___

162 one hundred sixty-two

Name _____

Subtract from 10

 Listen to the subtraction story. Use 🪙 to act it out.

10¢ − 5¢ = 5¢

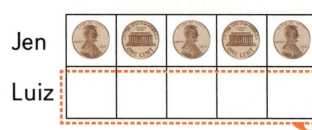
Jen
Luiz

Draw 🔴 to show the part taken away. Find the difference.

1.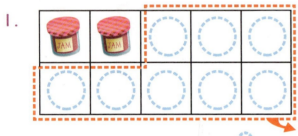

   10 − 8 = __2__

2.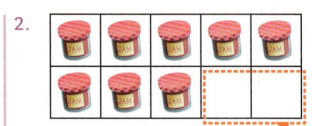

   10 − 2 = ___

3.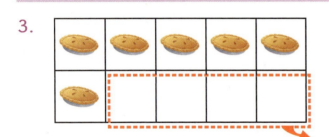

   10 − 4 = ___

4.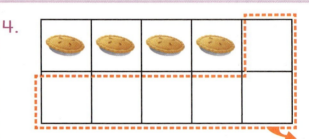

   10 − 6 = ___

5.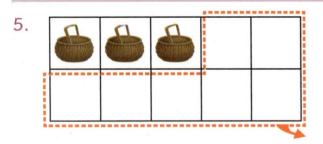

   10 − 7 = ___

6.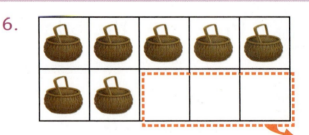

   10 − 3 = ___

Use 🪙 and a 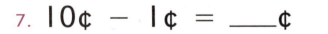. Subtract.

7. 10¢ − 1¢ = ___¢   10¢ − 5¢ = ___¢

8. 10¢ − 6¢ = ___¢   10¢ − 9¢ = ___¢

Have your child use 10 objects to retell each subtraction story shown above.

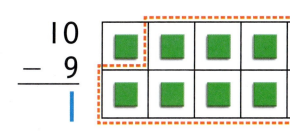
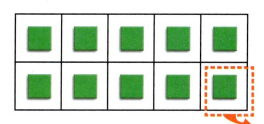

Both differences are **odd numbers**.

Subtract.

1. 10   9   10   8   9   10
   −7   −6   −5   −8   −5   −2
   ___ ___ ___ ___ ___ ___
    3

2. 7   9   9   10   8   9
   −2   −7   −1   −6   −4   −0
   ___ ___ ___ ___ ___ ___

3. 10   8   9   10   10   7
   −8   −6   −3   −9   −4   −1
   ___ ___ ___ ___ ___ ___

4. 9   10   8   9   10   7
   −2   −3   −2   −9   −1   −6
   ___ ___ ___ ___ ___ ___

 In 1–4 ✓ the odd differences.

5. Color 10 ☐ 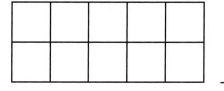.
   Color over 3 ☐ with blue. How many ☐ are left now? ___ ☐

6. Color an odd number of ☐. How many ☐ are not colored? ___ ☐

164 one hundred sixty-four

Subtract from 11

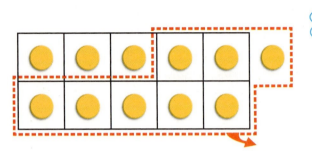

I know 10 − 8 = 2.
So 11 − 8 = 3.

11 − 8 = 3

Find the difference. Ring the ⬤ being taken away.

1.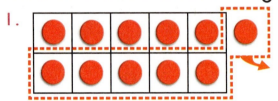

11 − 6 = 5 ←related facts→ 11 − 5 = ___

2.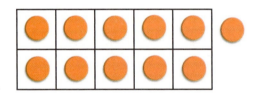

11 − 4 = ___ ←related facts→ 11 − ___ = ___

3.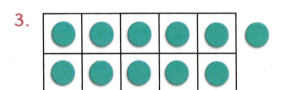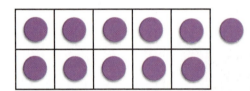

11 − 9 = ___ ←related facts→ 11 − ___ = ___

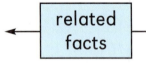 Subtract.
Match the 10 fact you used.

4. 11 − 5 = ___   11 − 3 = ___   11 − 2 = ___

10 − 3 = 7   10 − 5 = 5   10 − 2 = 8

 Make an 11-unit paper strip — for example, ▭▭▭▭▭▭▭▭▭▭▭ — and fold over some units. Have your child say the subtraction sentence.

one hundred sixty-five   165

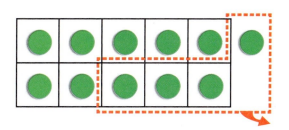
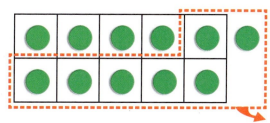

Subtract.

1. 11   10   9   11   11   8
   − 8 − 2 − 6 − 3 − 5 − 3
    3

2.   9   8   11   10   11   7
   − 2 − 6 − 9 − 6 − 4 − 7

3.   9   11   10   11   7   8
   − 8 − 6 − 3 − 7 − 1 − 4

4. 11   9   10   8   10   10
   − 2 − 4 − 8 − 7 − 5 − 7

 In 1–4 ring differences less than 5.

5. A clown holds 11 🎈.
   He gives 3 away.
   How many balloons
   does he have now?

   ___ ◯ ___ = ___

    now

6. There are 11 🤡.
   6 of them do a flip.
   How many clowns
   do not do a flip?

   ___ ◯ ___ = ___

Name _____

Subtract from 12

Amy has a dozen eggs.
6 are brown.
The rest are yellow.
How many eggs are yellow?

12 − 6 = 6

eggs in all / part taken away / part left

12
− 6
———
  6

Subtract. Ring the part taken away.

1.

   12 − 9 = 3   12 − 3 = ___

2.

   12 − 8 = ___   12 − 4 = ___

3.

   12 − 7 = ___   12 − 5 = ___

 In 1–3 use two colors to show related facts.

Find the differences. Fill in the table.

12 −	9	6	8	0	4	3	12	7	5

 Make a spinner that has 9 equally spaced sections numbered 1–9. Have your child take a spin and subtract the number from 12.

one hundred sixty-seven **167**

You can model or draw to check.

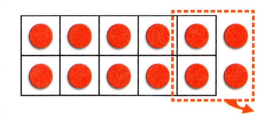

12
− 4
8

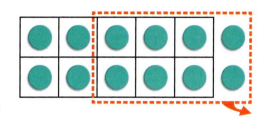

12
− 8
4

Subtract. Ring differences greater than 5.

1. 12   10   12   11   10   10
   − 7   − 4   − 8   − 7   − 1   − 5
    5

2. 11   9   12   11   12   9
   − 6   − 9   − 5   − 5   − 4   − 7

3. 11   11   8   11   10   12
   − 9   − 2   − 6   − 3   − 3   − 6

4.   9   12   8   12   11   9
   − 2   − 9   − 3   − 3   − 8   − 5

 Which problem has extra information?

5. Chris has 5¢.
   Linda has 10¢.
   Linda loses 7¢.
   How much does
   Linda have now?

   ____ ¢

6. Dan had a dozen 🥚.
   Some broke. Now he
   has 7 🥚. How many
   🥚 broke?

   ____ eggs

Name _____

Number-Line Subtraction

You can count back to subtract.

11 − 2 = __?__

Go to 11. Count back 2. Stop at 9.

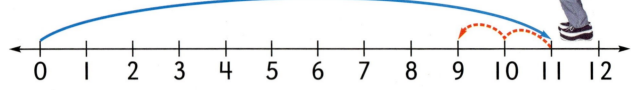

The number line shows 11 − 2 = 9.

Subtract. Show how you count back.

1. 12 − 5 = __7__

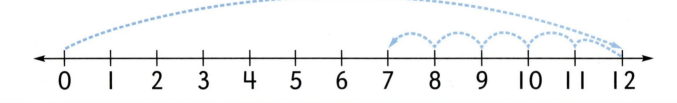

2. 10 − 2 = ____

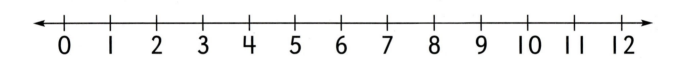

3. 12 − 4 = ____

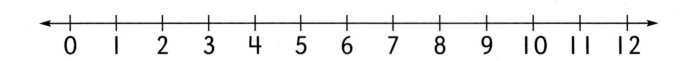

4. 11 − 3 = ____

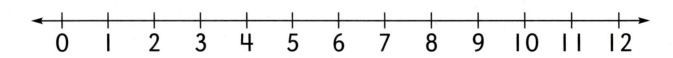

5-8 Have your child count back on a number line and write the corresponding subtraction sentence.

Write the subtraction sentence for each ⟵┈┈.

1.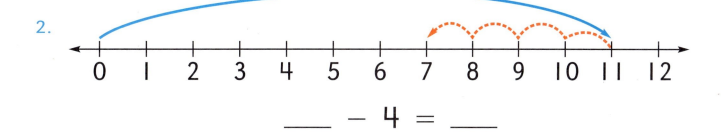

    12 − 7 = 5

2.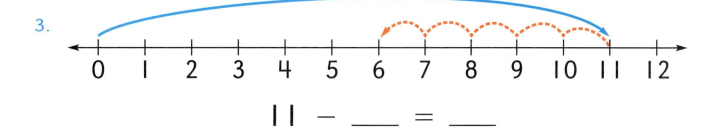

    ___ − 4 = ___

3.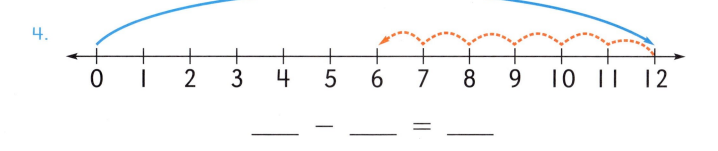

    11 − ___ = ___

4.

    ___ − ___ = ___

 FINDING TOGETHER Use a ⟵┈┈. Show 4 ways you can count back to subtract. Then write the number sentences.

5. ___ − ___ = ___   6. ___ − ___ = ___

7. ___ − ___ = ___   8. ___ − ___ = ___

170 one hundred seventy

Name _____

Subtract: Use Patterns

12 − 6 = 6
11 − 5 = 6
10 − 4 = 6

You can model subtraction patterns.

 How does the **number in all** change?
How does the **part taken away** change?
What happens to the **difference**?

Subtract. Look for the pattern.
Write the next number sentence.

1. 7 − 5 = 2
   8 − 6 = 2
   9 − 7 = 2
   10 − ___ = ___

2. 12 − 8 = ___
   11 − 7 = ___
   10 − 6 = ___
   ___ − ___ = ___

3. 6 − 4 = ___
   8 − 4 = ___
   10 − 4 = ___
   ___ − 4 = ___

4. 11 − 3 = ___
   9 − 3 = ___
   7 − 3 = ___
   ___ − 3 = ___

5. 12 − 9 = ___
   12 − 7 = ___
   12 − 5 = ___
   ___ − ___ = ___

6. 7 − 0 = ___
   7 − 2 = ___
   7 − 4 = ___
   ___ − ___ = ___

 Ask your child to tell how each of the patterns above was made. Then let him/her make up a pattern for you.

Subtract. Complete the pattern.

1.
```
   6       7       8       9       
  -5      -4      -3      -2      -__
  ___     ___     ___     ___     ___
   1       3
```

2.
```
  10      10      10      __      __
  -1      -3      -5      -__     -__
  ___     ___     ___     ___     ___
```

3.
```
   9       9       9       9       __
  -0      -2      -4      -6      -__
  ___     ___     ___     ___     ___
```

4.
```
   8       9      10      __      __
  -7      -6      -5      -__     -__
  ___     ___     ___     ___     ___
```

5. 11 − 2 = ___   6. 9 − 1 = ___

   11 − 5 = ___   8 − 1 = ___

   11 − 8 = ___   7 − 1 = ___

   ___ − ___ = 0   ___ − ___ = ___

Look for patterns.

 Make a pattern for each.

7. ___ − ___ = 1   8. ___ − ___ = 0

 9. Write and draw your favorite pattern.

172   one hundred seventy-two

Name _____

Subtraction Strategies

You can use these subtraction strategies.

Count back.

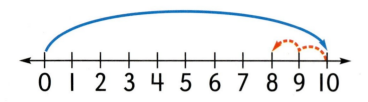

Look for patterns.

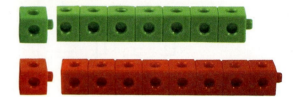

Take away.

Think 10.

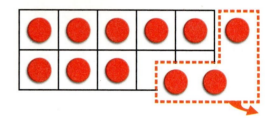

Find the difference. Which strategy did you use?

1. 11   12   11   11¢   10¢   11¢
   − 5   − 4   − 6   − 3¢   − 6¢   − 4¢
                       ¢   ¢   ¢

2. 12   10   12   12¢   10¢   12¢
   − 5   − 7   − 8   − 9¢   − 8¢   − 7¢
                       ¢   ¢   ¢

3. 11 − 7 = ___   11 − 8 = ___   11 − 9 = ___

4. 12 − 3 = ___   9 − 3 = ___   11 − 2 = ___

5. Subtract 0 from: 12, 7, 8, 11, 5, 9, 6.

6. Subtract 1 from: 10, 9, 6, 11, 8, 12, 7.

7. Tell what happens when you subtract 0 and 1.

Have your child describe how to use two of the strategies shown on this page to subtract.

one hundred seventy-three 173

Name _____

$2 + 5 = \underline{7}$

$4 + 4 = \underline{}$ $1 + 9 = \underline{}$

$7 + 2 = \underline{}$ $8 + 3 = \underline{}$ $6 + 6 = \underline{}$

$5 + 5 = \underline{}$ $0 + 9 = \underline{}$ $5 + 6 = \underline{}$

| 1 | 3 | 4 | 6 | 5 | 7 | 4 |
|+8 |+5 |+6 |+3 |+7 |+4 |+5 |

| 2 | 2 | 3 | 7 | 6 | 8 | 4 |
|+8 |+9 |+9 |+3 |+1 |+2 |+3 |

| 5 | 1 | 9 | 8 | 3 |
|+4 |+7 |+2 |+0 |+8 |

| 7 | 3 | 2 | 6 | 8 |
|+5 |+4 |+6 |+4 |+4 |

Color each sum.

7 red
9 blue
8 green

11 yellow
10 purple
12 orange

Name _____

Check by Adding

I know that 12 − 6 = 6 because 6 + 6 = 12.

To check 9 − 5 = 4 I use 4 + 5 = 9.

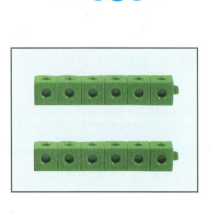

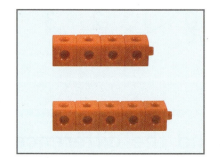

```
  12      6
−  6    + 6
────    ────
   6     12
```

These are related addition and subtraction facts.

```
   9      4
−  5    + 5
────    ────
   4      9
```

Subtract. Check by adding. You can use .

1.
```
  10       5      11              10
−  5    + 5     −  4    +       −  8    +
────    ────    ────   ────     ────   ────
```

2.
```
   7            12               8
−  3    +     −  8    +        − 3    +
────   ────   ────   ────      ────   ────
```

3.
```
  12            9               11
−  9    +     − 6    +        −  5    +
────   ────   ────   ────     ────   ────
```

 TALK IT OVER When did you use doubles? Where are the parts and the whole in each fact in 1 and 3?

5-11 Give your child several subtraction sentences—with a few mistakes. Ask your child to check your subtraction by adding.

one hundred seventy-five **175**

You can show related addition and subtraction facts on a number line.

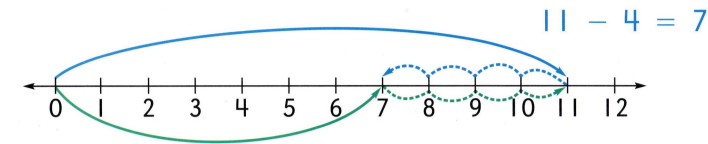

$11 - 4 = 7$

$7 + 4 = 11$

Find the difference. Check on a ↔.

1. 12   11   8   10   12   11
   − 3   − 6 − 0   − 9 − 4   − 5

2. 12   7   12   9   7   11
   − 6 − 2   − 7 − 9 − 5 − 3

3. 9   10   7   11   9   12
   − 2 − 7 − 4 − 9 − 0 − 5

 Write the related number sentences.

4. The difference is 5.
   I am a doubles fact.
   What am I?

   ___ − ___ = ___

   ___ + ___ = ___

5. The sum is 8¢.
   I am a doubles fact.
   What am I?

   ___¢ + ___¢ = ___¢

   ___¢ − ___¢ = ___¢

Name _____

PROBLEM-SOLVING STRATEGY
Draw to Compare

1. **Read** Pedro has 8 .
   Maria has 6 .
   Who has more ?
   How many more?

   To find how many more, subtract.

   Draw Pedro | ○○○○○○○○
   Maria | ○○○○○○

   Think Who has more ?   (Pedro)   Maria

   Write _8_ − _6_ = _2_

   Pedro has _2_ more .

2. **Read** Jan used 9 .
   Bill used 5 .
   Who has used fewer ?
   How many fewer?

   To find how many fewer, subtract.

   Draw Jan |
   Bill |

   Think Who used fewer ?   Bill   Jan

   Write ___ ○ ___ = ___

   _____ used ___ fewer .

5-13 Put 12 pennies on a table. Have your child take some; you take the rest. Ask your child to find how many more/fewer you have.

one hundred seventy-nine   **179**

Use these steps. Read → Draw → Think → Write

1. Juan had 3 🎈.
   Anna had 7 🎈.
   Who had more 🎈?
   How many more?

   Juan
   ―――――
   Anna

   ___ ◯ ___ = ___   _____ had ___ more 🎈.

2. Miki has 12¢.
   Keisha has 5¢.
   Who has more 🪙?
   How much more?

   Miki
   ―――――
   Keisha

   ___¢ ◯ ___¢ = ___¢   _____ has ___¢ more.

3. Alicia won 10 🎗.
   Lynn won 7 🎗.
   Who won fewer 🎗?
   How many fewer?

   Alicia
   ―――――
   Lynn

   ___ ◯ ___ = ___   _____ won ___ fewer 🎗.

MAKE UP YOUR OWN

4. Carl spent ___ 🪙.
   Leah spent ___¢.
   _____ spent ___¢ less.

5. Mia lost ___¢.
   Ross lost ___¢.
   _____ lost ___¢ more.

PROBLEM SOLVING

180   one hundred eighty

Name _____

PROBLEM-SOLVING APPLICATIONS
Use a Model

Read → Model → Think → Write

1. I saw 11 .
   6 of them had spots.
   How many did not have spots?

   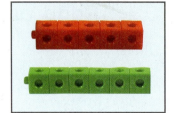

   add
   (subtract)

   11
   − 6

   ____ did not have spots.

2. Pedro saw 5 .
   Then Li saw 2 more.
   How many did they see in all?

   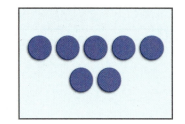

   (add)
   subtract

   5
   + 2

   They saw ____ in all.

3. Jamal catches 8 .
   2  hop away.
   How many are left?

   add
   subtract

   ____  are left.

4. Sue makes a dozen 🥧.
   8 are cherry. How many 🥧 are not cherry?

   add
   subtract

   ____ 🥧 are not cherry.

5. Ruth catches 4 .
   She catches 6 more.
   How many does she have now?

   add
   subtract

   She has ____ now.

PROBLEM SOLVING

5-14 Have your child use counters to model problems for addition and subtraction facts to 12.

one hundred eighty-one 181

Use a strategy you have learned.

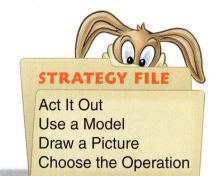

STRATEGY FILE
Act It Out
Use a Model
Draw a Picture
Choose the Operation

1. You see 11 🐥.
   9 of them fly away.
   How many 🐥 are left?

   add or subtract

   ___ ◯ ___ = ___

   ___ 🐥 are left.

2. I have 10 🪙.
   I spend all of them.
   How much do I have now?

   add or subtract

   ___ ◯ ___ = ___

   I have ___ 🪙 now.

3. There are 7 🎈 in the sky.
   2 of them come down.
   How many 🎈 are in the sky now?

   add or subtract

   ___ ◯ ___ = ___

   ___ 🎈 are in the sky now.

4. You find 6 🪙.
   You find 3¢ more.
   How much do you have now?

   add or subtract

   ___¢ ◯ ___¢ = ___¢

   You have ___¢ now.

5. ___ 🦆 are in a pond.

   ___ 🦆 swim away   or   come.

   How many 🦆 are

   left   or   in all?   ___ ◯ ___ = ___

Name _____

Chapter Review and Practice

Subtract. Show your work on a ⟵┼┼┼┼⟶.

1. $8 - 2 = \underline{6}$ $12 - 7 = \underline{}$ $10 - 7 = \underline{}$
2. $9 - 8 = \underline{}$ $7 - 4 = \underline{}$ $10 - 9 = \underline{}$
3. $9 - 7 = \underline{}$ $10 - 8 = \underline{}$ $11 - 4 = \underline{}$
4. $12 - 3 = \underline{}$ $9 - 3 = \underline{}$ $11 - 2 = \underline{}$

Find the difference. Check by adding.

5. 9   5   7   11   12¢
   −4 +4   −5   − 7   − 9¢
   ─── ── ─── ──── ─────
    5

6. 12   4   9   8   9¢
   − 8 +8   −6   −4   −5¢
   ─── ── ───   ───   ─────

7. 11   8   12   8   10¢
   − 3 +3 − 7   −6   − 6¢
   ─── ── ───   ───   ─────

8.   8   3   11   7   10¢
   − 5 +5 − 9   −3   − 3¢
   ─── ── ───   ───   ─────

PROBLEM SOLVING

9. There are 12 🐟.
   5 swim away.
   How many are left?

   _____

10. Li has 10 🪙.
    Tuan has 5¢.
    Who has more?
    How much more?

    _____

This page reviews the mathematical content presented in Chapter 5.

Name _____

Use a . Spin for odd or even.
Fill in each table.

Write the greater number first.

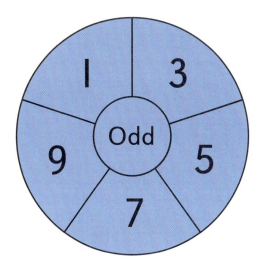
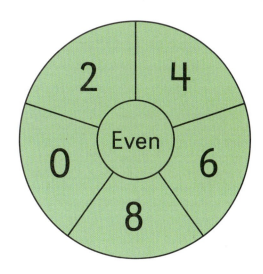

Odd + Odd =	Odd	Even

Odd − Odd =	Odd	Even

Odd + Even =	Odd	Even

Odd − Even =	Odd	Even

Even + Even =	Odd	Even

Even − Even =	Odd	Even

What patterns did you find?
Predict the answers to:
Even + Odd = ?   Even − Odd = ?

Performance Assessment

1. Use 🟩 or 🟦.
   Show this pattern.
   Then write the next fact.

11	11	11	
−11	− 9	− 7	− __
___	___	___	___

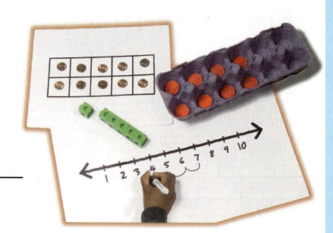

2. Draw 🔴 to take away from 8.
   Complete. Write the related subtraction sentence.

   8 − 3 = ___   ___ − ___ = ___

3. Use the ⇠┈┈⇢.
   Write the sentence.
   Check by adding.

   ___ ◯ ___ = ___

   ___ + ___ = ___

 PORTFOLIO Choose 1 of these projects. Use a separate sheet of paper.

4. Show 2 fact families that have 4 sentences and 2 that do not. (Hint: Think about doubles.)

5. Make a chart like this. List the facts for 10 you use to subtract from 9 and 11.

Subtract from 9	Fact for 10	Subtract from 11
9 − 1 = 8	10 − 1 = 9	11 − 1 = 10

This page provides a variety of informal assessment opportunities in order to measure your child's understanding of the skills taught in Chapter 5.

one hundred eighty-five **185**

Check Your Mastery

Name _____

Find the difference.

1.   9   7   8   11   10¢   12¢
    −7   −4   −6   − 8   − 4¢   − 7¢
                                                        ¢   ¢

2. 10   11   12   9   7¢   11¢
    − 3   − 6   − 4   −3   −2¢   − 2¢
                                                        ¢   ¢

3.   9   12   10   8   10¢   9¢
    −5   − 3   − 2   −3   − 6¢   −2¢
                                                        ¢   ¢

Subtract.

4. 8 − 4 = ___   10 − 7 = ___   11 − 3 = ___

5. 11 − 7 = ___   12 − 9 = ___   7 − 5 = ___

6. 10 − 8 = ___   8 − 5 = ___   9 − 6 = ___

Use a ↔. Complete each pattern.

7. 8 − 7 = ___ | 10 − 1 = ___   7 − 7 = ___

    9 − 8 = ___ | 9 − 1 = ___   8 − 8 = ___

    10 − ___ = ___ | ___ − ___ = ___   ___ − ___ = ___

PROBLEM SOLVING Use a strategy you have learned.

8. Lara has 6 .
   She finds 6¢ more.
   How much does
   she have now? _____ ¢

9. Jon drew a dozen △.
   He colored 8 blue.
   How many are
   not blue? _____

186 one hundred eighty-six

This page is a formal assessment of your child's understanding of the content presented in Chapter 5.

Name _____

Cumulative Test 1
Chapters 1–5

Mark the ○ for your answer.

Listening Section

A

8	6	7
○	○	○

B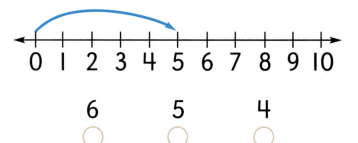

6	5	4
○	○	○

1. Add.

2 + 5 = ___

6	7	8	9
○	○	○	○

2. Subtract.

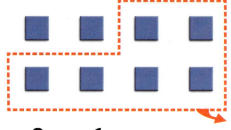

8 − 6 = ___

3	6	4	2
○	○	○	○

3. Find the sum.

   6
+4

8	2	10	9
○	○	○	○

4. Find the difference.

  11
− 6

5	3	6	2
○	○	○	○

5. Find the related subtraction fact.

  7
+2
  9

○ 7 − 2 = 5
○ 9 − 2 = 7
○ 2 + 9 = 11
○ 5 + 2 = 7

6. Find the number sentence.

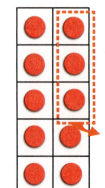

○ 10 − 7 = 3
○ 9 − 3 = 7
○ 10 − 3 = 7
○ 7 + 2 = 9

one hundred eighty-seven 187

Mark the ○ for your answer.

7. Find the missing number.

   8 − 8 = □
   7 + □ = 7

   - ○ 0
   - ○ 1
   - ○ 12
   - ○ 2

8. What fact is next in the pattern?

   1 + 1   3 + 3   ? + ?

   - ○ 2 + 2
   - ○ 4 + 4
   - ○ 5 + 5
   - ○ 4 + 5

9. Find sums of 7.

   3 + 4 = 2 + ___

   - ○ 6
   - ○ 2
   - ○ 1
   - ○ 5

10. Find the missing number.

    7 − □ = 1

    - ○ 1
    - ○ 6
    - ○ 7
    - ○ 8

11. 9¢ + 3¢ = ___

    - ○ 6¢
    - ○ 10¢
    - ○ 11¢
    - ○ 12¢

12. 11¢ − 9¢ = ___

    - ○ 10¢
    - ○ 2¢
    - ○ 8¢
    - ○ 3¢

13. There are 3 .
    6 more come. How many are there now?

    - ○ 10
    - ○ 9
    - ○ 3
    - ○ 8

14. You have 7 ¢.
    You find 5¢ more. How much do you have now?

    - ○ 2¢
    - ○ 3¢
    - ○ 12¢
    - ○ 11¢

15. Rosa has 11¢.
    Tina has 8¢. How much more does Rosa have?

    ___ ○ ___ = ___

    - ○ 4¢
    - ○ 2¢
    - ○ 5¢
    - ○ 3¢

16. You draw a dozen □.
    You color 4 blue. How many are not blue?

    ___ ○ ___ = ___

    - ○ 8
    - ○ 9
    - ○ 4
    - ○ 7

Place Value to 100

6

CRITICAL THINKING

44 is 4 groups of ten and 4.
33 is 3 groups of ten and 3.
What is 22? Find it in the picture.

For more information about Chapter 6, visit the Family Information Center at www.sadlier-oxford.com

Dear Family,

Today your child began Chapter 6. As she/he studies place value to 100, you may want to read the poem below, which was read in class. Encourage your child to talk about some of the math ideas shown on page 189.

The 🏠 at the bottom of each skills lesson suggests ways for you to help your child. You may want to have number cards (1 to 100) and countables available for your child to use throughout this chapter.

Home Activity

Place-Value Scramble

Try this activity with your child. Place number cards (1–10) in a bag. Shake the bag to scramble the cards; then empty it so that the number cards are strewn on the floor or table. Give your child about two minutes to put the cards in order. After he/she finishes each lesson in this chapter, adjust the range and number of cards used.

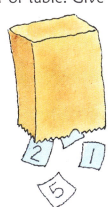

Home Reading Connection

Math My Way

Two plus two is twenty-two.
It's plain as day that this is true.
But teacher says she's very sure
That two plus two adds up to four.

Three plus three makes thirty-three.
That's the way it ought to be.
But teacher says the answer's six.
I don't know why. Must be a trick.

Four plus four is forty-four.
Not any less, not any more.
My teacher just can't get it straight.
She keeps on saying the answer's eight.

I give up. I'll go along.
I'll do it her way, though she's wrong.
But in my heart, I know what's true—
Two plus two makes twenty-two.

Carol Diggory Shields

Name _____

MATH CONNECTIONS
Making Ten

10 → 1 ten   11 → 1 ten 1 one   12 → 1 ten 2 ones

Count each group of toys.
Write the number.

____   ____   ____   ____   ____   ____   ____

 TALK IT OVER Tell how many tens and ones for each number above.
Which group of toys has less than 10 ones?
How many more toys does it need to make 1 ten?

 PORTFOLIO You can put this in your Math Portfolio.

one hundred ninety-one **191**

CROSS-CURRICULAR CONNECTIONS
Multicultural Studies

Name _____

Ten is important in any language.

Ring groups of 10.

1. Japanese juu

2. Spanish diez

3. Filipino sampú

4. French dix

5. German zehn

6. Write the number of toys left over.

 10 and ____   10 and ____   10 and ____

 10 and ____   10 and ____

Name _____

Numbers 10–19

tens	ones

place value

1 ten 3 ones
10 + 3
13
thirteen

In a 2-digit number each place has a special value.

Write how many.

1. 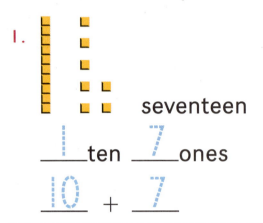 seventeen

   _1_ ten _7_ ones

   _10_ + _7_   _17_

2. fourteen

   ___ ten ___ ones

   ___ + ___   ___

3. 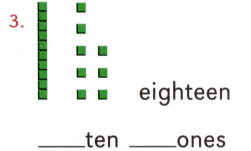 eighteen

   ___ ten ___ ones

   ___ + ___   ___

4.  sixteen

   ___ ten ___ ones

   ___ + ___   ___

5. fifteen

   ___ ten ___ ones

   ___ + ___   ___

6. 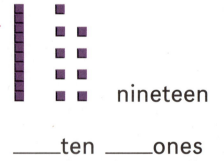 nineteen

   ___ ten ___ ones

   ___ + ___   ___

 TALK IT OVER Which is the least number on the page?

Help your child look through newspapers, magazines, and catalogs for numbers from 10 to 19.

one hundred ninety-three **193**

Draw to finish the pattern. Write the missing tens and ones.

1. 11 __1__ ten __1__ one   12 __1__ ten __2__ ones   13 __1__ ten ____ ones

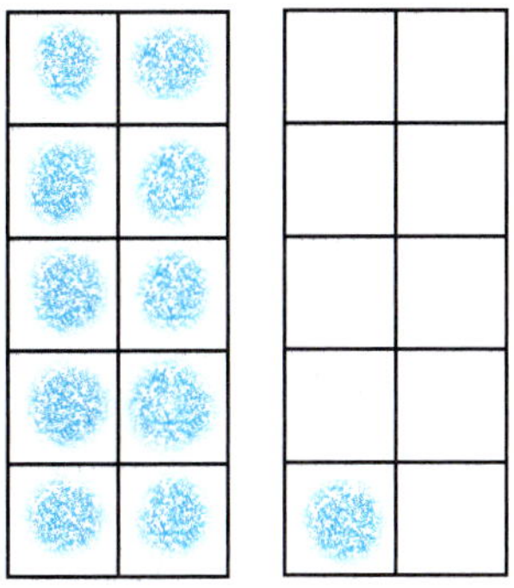

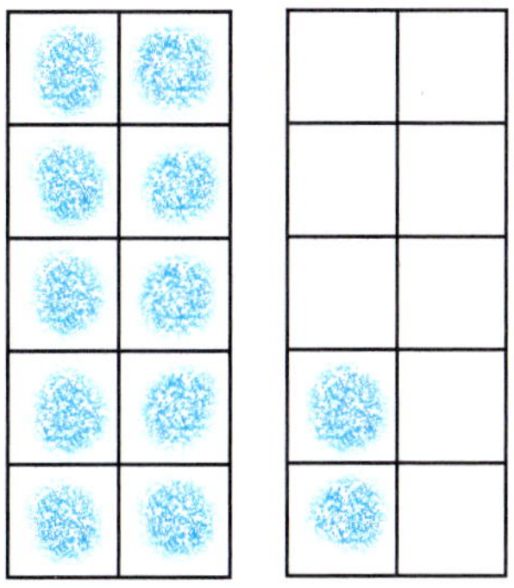

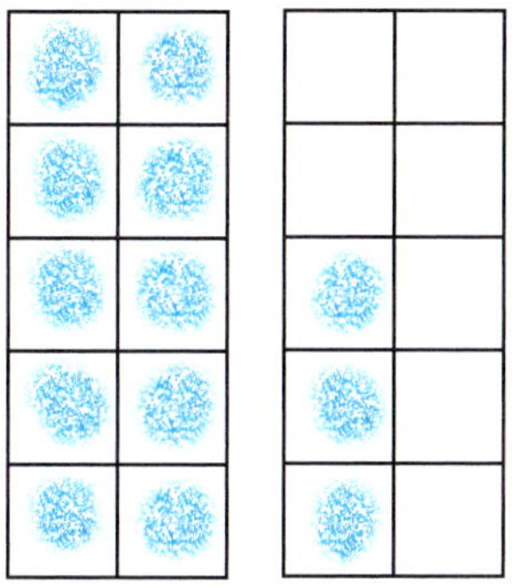

2. 14 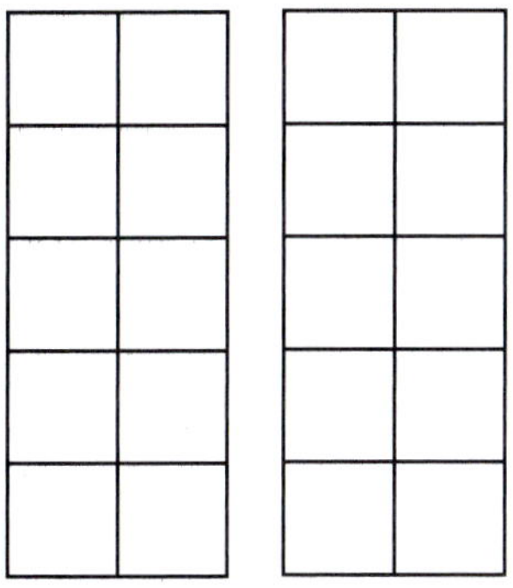 ____ ten ____ ones   15 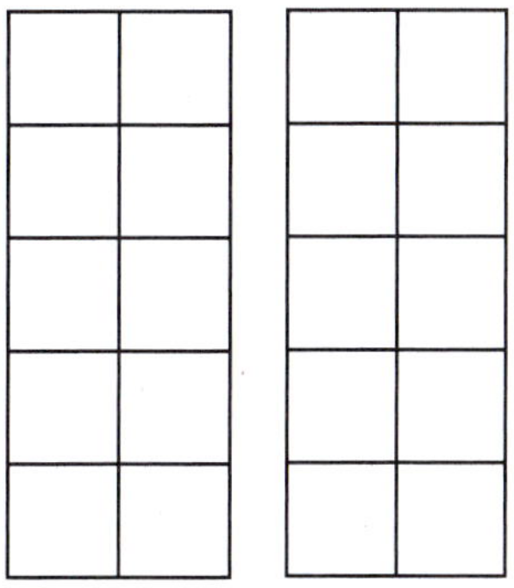 ____ ten ____ ones   16  ____ ten ____ ones

3. 17 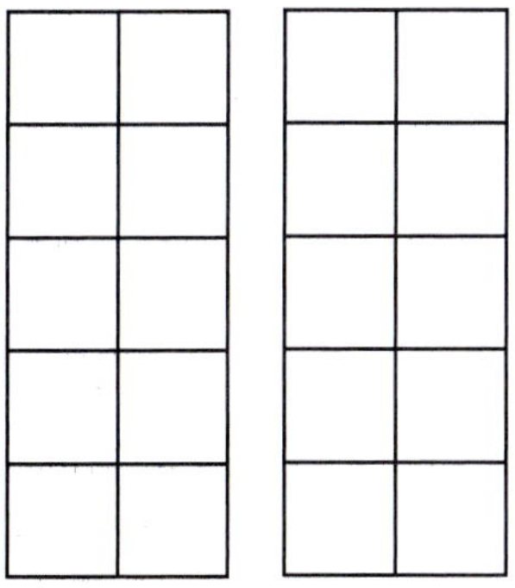 ____ ten ____ ones   18  ____ ten ____ ones   19 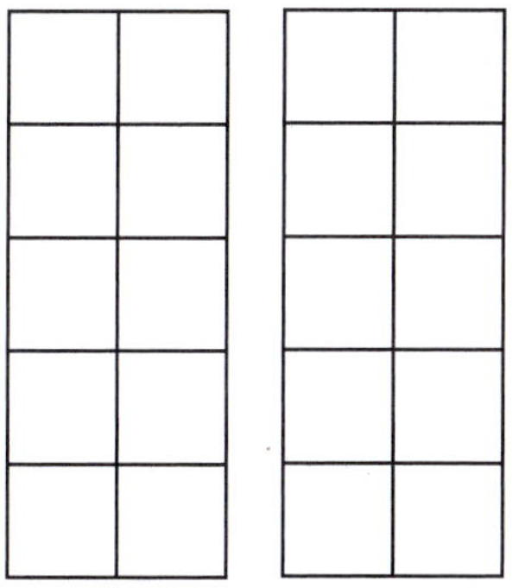 ____ ten ____ ones

PROBLEM SOLVING

has 1 ten 6 ones.  has ▭▭▭ ▫▫▫▫▫.

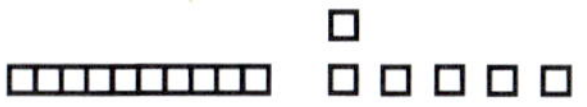

has 13. Ring the answer.

4.  has more or fewer than  .

5.  and  have different or same number.

194 one hundred ninety-four

Name _____

Ten to Nineteen

15 fifteen

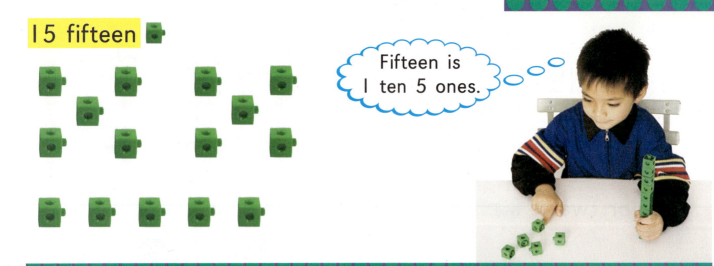

Fifteen is 1 ten 5 ones.

Write the number and the number word. Color the number.

1. 1 ten 8 ones __18__

   eighteen

2. 1 ten 2 ones ____

   ____

3. 1 ten 7 ones ____

   ____

4. 1 ten 4 ones ____

   ____

5. 1 ten 9 ones ____

   ____

6. 1 ten 3 ones ____

   ____

7. Write the number and the number word for 1 ten 6 ones. ____ ____

Describe a number from 10–19 — for example, "My number has 1 ten 4 ones" — and have your child name and write it.

A number line shows the **order** of numbers.

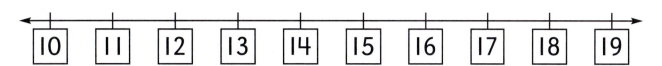

1. Color thirteen red.
2. Color fifteen blue.
3. Color twelve green.
4. Color eleven purple.

5. ✔ the number with 1 ten 6 ones.

6. ✗ the number with 1 ten 9 ones.

7. ◯ the number with 1 ten 1 one.

Write the missing number.

8. 10, 11, 12, _13_
9. 14, 15, 16, ___
10. 11, 12, 13, ___
11. 8, 9, 10, ___
12. 16, 17, 18, ___
13. 13, 14, 15, ___
14. 9, 10, 11, ___
15. 15, 16, 17, ___

 16. Write each number from 10 to 19 in 3 different ways.

| 10 | ten | 1 ten 0 ones |

Count by Tens

Name _____

You can use models to count by tens.

1 ten	2 tens	3 tens	4 tens	5 tens
10	20	30	40	50
ten	twenty	thirty	forty	fifty

6 tens	7 tens	8 tens	9 tens	10 tens
60	70	80	90	100
sixty	seventy	eighty	ninety	one hundred

Talk It Over — Tell how to model the numbers 60, 70, 80, and 90.

Write how many tens.
Write the number and number word.

1. __2__ tens __20__
   twenty

2. ____ tens ____
   forty

3.  ____ tens ____
   sixty

4. 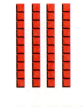 ____ tens ____
   eighty

5. 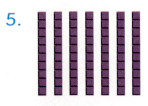 ____ tens ____
   seventy

6. 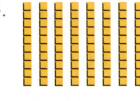 ____ tens ____
   fifty

7. ____ tens ____
   thirty

8. 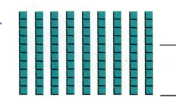 ____ tens ____
   ninety

Have your child arrange number word cards for 10, 20, 30, ..., 90 in order and write each number.

one hundred ninety-seven 197

You can count on by tens on a number line.

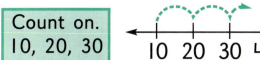 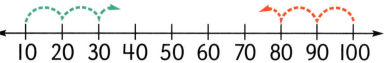

 How is counting by tens like counting by ones? How is it different?

Write the missing numbers. Use a number line.

1. 10, 20, __30__, ___, 50, ___, 70, ___, 90

2. 20, ___, 40, ___, ___, 70, 80, ___, 100

3. 90, 80, ___, ___, 50, 40, 30, ___, ___

4. 100, 90, ___, 70, ___, ___, 40, ___, 20

Write the number.

5. 2 tens __20__   forty ___   6 tens ___

6. ninety ___   7 tens ___   fifty ___

7. I have 20 ✶.
I get ten more.
Now I have ___ ✶.

8. I have forty .
I give 10 away.
Now I have ___ .

9. What is 2 tens fewer than ninety? ___

10. What is 2 tens more than fifty? ___

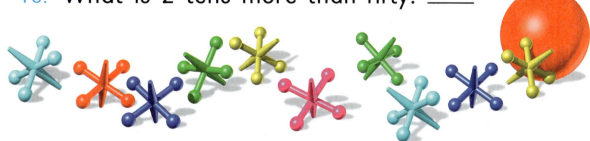

198 one hundred ninety-eight

Name _____

Numbers 20–39

A **place-value mat** shows tens and ones.

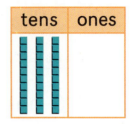
2 tens 9 ones
29 twenty-nine

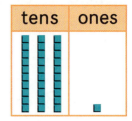
3 tens 0 ones
30 thirty

3 tens 1 one
31 thirty-one

Write how many.

1. ___2___ tens ___6___ ones
   ___26___ twenty-six

2.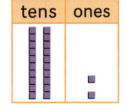
   _____ tens _____ ones
   _____ twenty-two

3.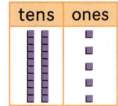
   _____ tens _____ ones
   _____ twenty-five

4.
   _____ tens _____ ones
   _____ thirty-three

5.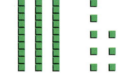
   _____ tens _____ ones
   _____ thirty-eight

6.
   _____ tens _____ ones
   _____ thirty-four

7.
   _____ tens _____ one
   _____ twenty-one

8.
   _____ tens _____ ones
   _____ thirty-nine

9.
   _____ tens _____ ones
   _____ thirty-seven

 How are 23 and 32 different?
Which is 1 more than 31?

 6-4 Describe a number between 20 and 39—for example, "My number has 3 tens and 4 ones." Have your child model and name the number.

one hundred ninety-nine 199

Model each. Write the number.

1. 2 tens 7 ones	2 tens 5 ones	2 tens 3 ones
27	____	____
2. 3 tens 8 ones	3 tens 6 ones	3 tens 4 ones
____	____	____

Model each. Write the number word.

3. 2 tens 0 ones	2 tens 4 ones	2 tens 8 ones
twenty	twenty-____	twenty-____
4. 3 tens 1 one	3 tens 2 ones	3 tens 3 ones
thirty-____	thirty-____	thirty-____

 Describe the patterns in 1–4.

Count by 1s. Write the missing numbers.

5.

11	12	13	14	15	___	17	18	___	20
21	22	___	24	___	26	___	___	29	___
___	___	33	___	35	___	37	___	___	___

 6. What comes next?

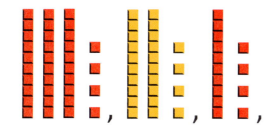

____ tens ____ ones

Name _____

Numbers 40–59

tens	ones
||||| (4 tens)	⋮ (9 ones)

4 tens 9 ones
49 forty-nine

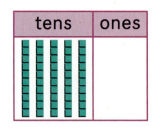

5 tens 0 ones
50 fifty

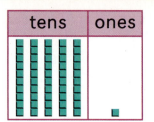

5 tens 1 one
51 fifty-one

Write how many.

1. [tens ||||| ones ⋮⋮]
   _5_ tens _4_ ones
   _54_ fifty-four

2.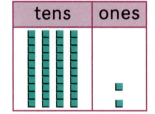
   ___ tens ___ ones
   ___ forty-two

3.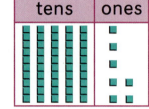
   ___ tens ___ ones
   ___ fifty-seven

4.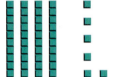
   ___ tens ___ ones
   ___ forty-six

5.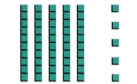
   ___ tens ___ ones
   ___ fifty-five

6.
   ___ tens ___ ones
   ___ forty-eight

7.
   ___ tens ___ ones
   ___ fifty-nine

8.
   ___ tens ___ ones
   ___ forty-five

9.
   ___ tens ___ ones
   ___ fifty-eight

 TALK IT OVER Which has more tens, 47 or 56?

 6-5 Ask your child to name numbers between 0 and 59 that have the same number of tens and ones—for example, 11 and 22.

two hundred one **201**

Write the number or number word. Model each.

1. 4 tens 7 ones	4 tens 8 ones	4 tens 9 ones
47	___	___
2. 5 tens 4 ones	5 tens 3 ones	5 tens 2 ones
___ fifty-___	___ fifty-___	___ fifty-___
3. 4 tens 0 ones	4 tens 3 ones	4 tens 6 ones
forty	forty-___	forty-___
4. 5 tens 5 ones	4 tens 4 ones	3 tens 3 ones
___ fifty-___	___ forty-___	___ thirty-___

 Describe the pattern in each row above. What number comes next in 1–4?

Count by 1s. Write the missing numbers.

5.

41	42	43	___	45	___	___	48	___	50
51	52	___	___	___	___	57	___	___	

 Use ▭▭▭▭ and ▪. How are these numbers different? Ring.

6. Which is more?   7. Which is most?

42   or   52   53   54   55

Name _____

Numbers through 59

Color to match.

1.

Number Word	Number	Place Value	Model
twenty-seven	46	5 tens 2 ones	
fifty-two	27	4 tens 6 ones	
forty-six	52	2 tens 7 ones	
fifty-three	34	3 tens 4 ones	
thirty-four	53	5 tens 3 ones	

Write the place value and number.

2. forty-seven 47
   4 tens 7 ones

3. twenty-six ____
   ___ tens ___ ones

4. fifteen ____
   ___ ten ___ ones

5. thirty-eight ____
   ___ tens ___ ones

6. fifty-one ____
   ___ tens ___ one

7. forty-nine ____
   ___ tens ___ ones

8. thirteen ____
   ___ ten ___ ones

9. forty-five ____
   ___ tens ___ ones

10. twenty-three ____
    ___ tens ___ ones

 Are 52 and 25 the same? Which has fewer tens? Which has fewer ones?

6-6 Say a number from 1 to 58. Then have your child count on to 59.

two hundred three 203

You can show the numbers to 59 on a number line.

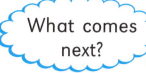
What comes next?

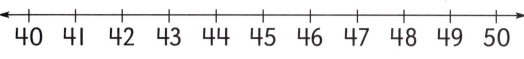

Write the missing numbers.

1.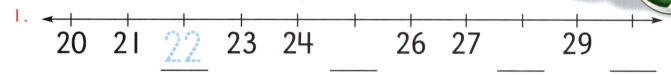
20 21 22 23 24 ___ 26 27 ___ 29 ___

2.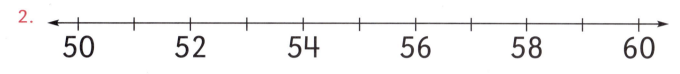
50 ___ 52 ___ 54 ___ 56 ___ 58 ___ 60

3.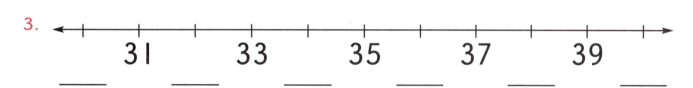
___ 31 ___ 33 ___ 35 ___ 37 ___ 39 ___

4. Connect the dots in order. Begin at 10.

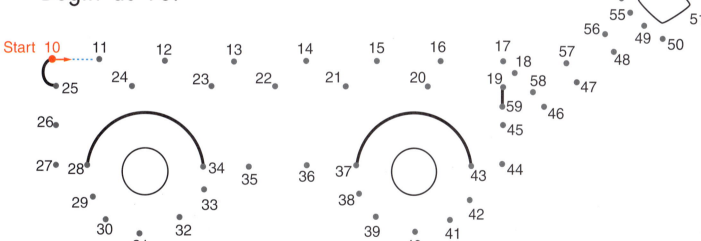

 CRITICAL THINKING

Find the missing number.

5. 5 tens + ___ = 50

6. 4 ones − 4 ones = ___

7. ___ + 3 tens = 30

8. 2 ones − ___ = 2 ones

Name _____

Numbers 60–79

You can model numbers on a place-value mat.

tens	ones

6 tens 9 ones
69 sixty-nine

7 tens 0 ones
70 seventy

7 tens 1 one
71 seventy-one

Write how many.

1.
   _7_ tens _2_ ones
   _72_ seventy-two

2. ___ tens ___ ones
   ___ sixty

3.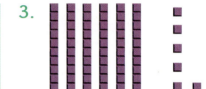
   ___ tens ___ ones
   ___ sixty-six

4.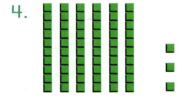
   ___ tens ___ ones
   ___ sixty-three

5.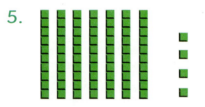
   ___ tens ___ ones
   ___ seventy-four

6.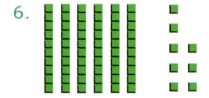
   ___ tens ___ ones
   ___ sixty-eight

7.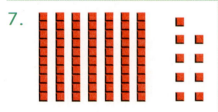
   ___ tens ___ ones
   ___ seventy-nine

8.
   ___ tens ___ ones
   ___ sixty-seven

9.
   ___ tens ___ ones
   ___ seventy-seven

Work with your child to list the ages of family members. Then have her/him model the ages with countables.

Write the number or number word. Model each.

1. 7 tens 3 ones	7 tens 5 ones	7 tens 7 ones
_73_	____	____
2. 6 tens 2 ones	6 tens 4 ones	6 tens 6 ones
____	____	____
3. 7 tens 5 ones	6 tens 5 ones	6 tens 1 one
seventy-_five_	sixty-____	sixty-____
4. 7 tens 6 ones	6 tens 9 ones	7 tens 8 ones
seventy-____	sixty-____	seventy-____

 Which rows above make a pattern?

5. Count by 1s. Write the missing numbers.

61	62	63	___	___	___	___	69	___
71	___	___	74	___	76	___	78	___	80

 6. Sixty-two is an even number. Find other even numbers in 5. Color these even numbers yellow.

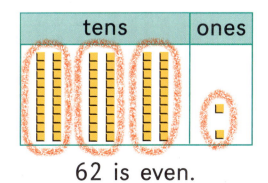

62 is even.

206 two hundred six

Name _____

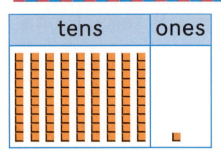

tens	ones
8 tens 9 ones	
89 eighty-nine	

tens	ones
9 tens 0 ones	
90 ninety	

tens	ones
9 tens 1 one	
91 ninety-one	

Write how many.

1.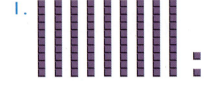
   _9_ tens _2_ ones
   _92_ ninety-two

2.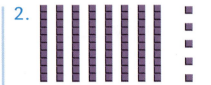
   ___ tens ___ ones
   ___ eighty-five

3.
   ___ tens ___ ones
   ___ eighty-seven

4.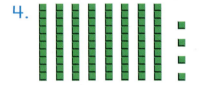
   ___ tens ___ ones
   ___ eighty-four

5.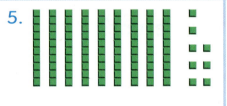
   ___ tens ___ ones
   ___ ninety-eight

6.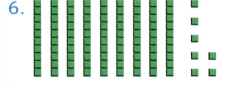
   ___ tens ___ ones
   ___ ninety-seven

7.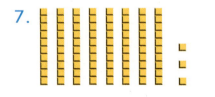
   ___ tens ___ ones
   ___ eighty-three

8.
   ___ tens ___ ones
   ___ ninety-nine

9.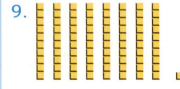
   ___ tens ___ one
   ___ eighty-one

TALK IT OVER Name the numbers that have 8 tens.
Which number in 1–9 has 8 ones?

Write the number or number word. Model each.

1. 9 tens 1 one   9 tens 5 ones   9 tens 3 ones

   91   _____   _____

2. 8 tens 0 ones   8 tens 2 ones   8 tens 8 ones

   _____   _____   _____

3. 8 tens 6 ones   9 tens 6 ones   7 tens 6 ones

   eighty-_____   ninety-_____   seventy-_____

4. 9 tens 4 ones   6 tens 4 ones   8 tens 4 ones

   ninety-_____   sixty-_____   eighty-_____

 SHARE YOUR THINKING How are the numbers in each row the same? How are they different?

Complete the place value.

5. 97 = **9** tens **7** ones   6. 93 = ___ tens ___ ones

7. 89 = ___ tens ___ ones   8. 64 = ___ tens ___ ones

9. 58 = ___ tens ___ ones   10. 90 = ___ tens ___ ones

11. 77 = ___ tens ___ ones   12. 91 = ___ tens ___ one

 CHALLENGE

13. Write each number. ✔ the odd numbers.

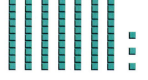      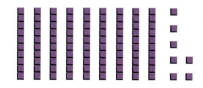

_____   _____   _____

Name _____

Numbers to 100

You can make a number line for each group of numbers.

90 to 100
90 91 92 93 94 95 96 97 98 99 100

Write the missing numbers.

1. **80 to 90**
80 81 82 83 ___ 85 86 ___ 88 89 ___

2. **70 to 80**
70 71 ___ 73 ___ 75 ___ 77 78 ___ 80

3. **60 to 70**
___ 61 ___ 63 ___ 65 66 67 ___ 69 70

4. **50 to 60**
50 ___ 52 ___ 54 ___ 56 ___ 58 ___ 60

Share Your Thinking: How are these number lines the same? How are they different?

5. Put these numbers in order.

71 ___ ___ ___ ___

Boxes: 80, 71, 76, 74, 78

Have your child list as many numbers with 9s in them as he/she can.

Complete the place value.
Color the number.

1. thirty-five ones

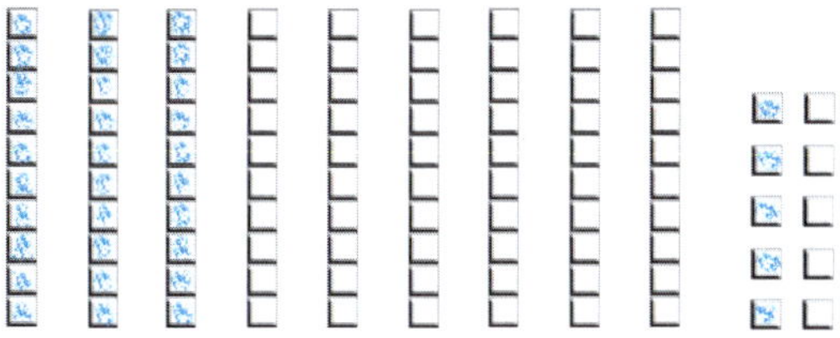

   _35_ ones = _3_ tens _5_ ones

   fifty-three ones

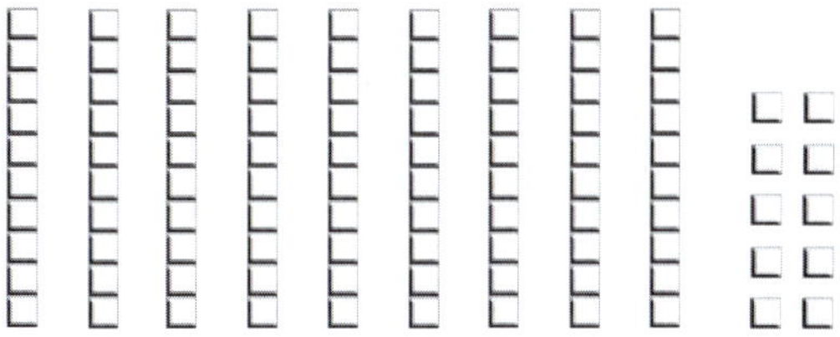

   ___ones = ___tens ___ones

2. eighty-nine ones

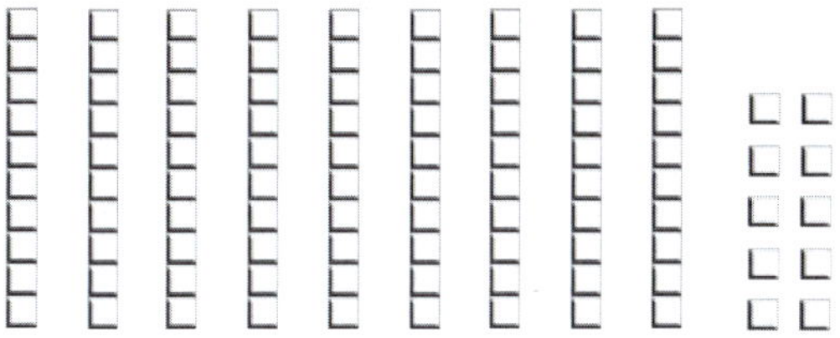

   ___ones = ___tens ___ones

   ninety-eight ones

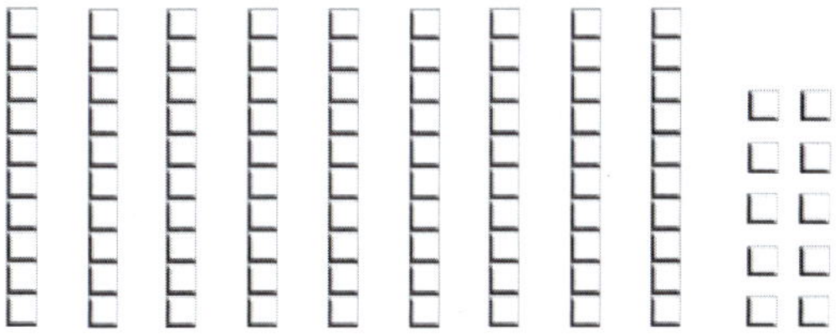

   ___ones = ___tens ___ones

3. seventy-nine ones

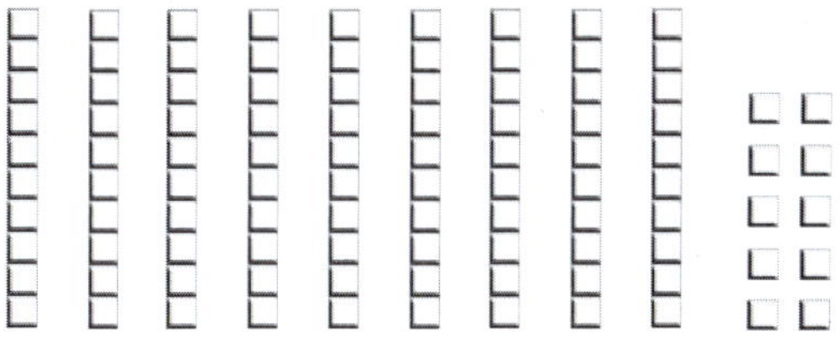

   ___ones = ___tens ___ones

   ninety-seven ones

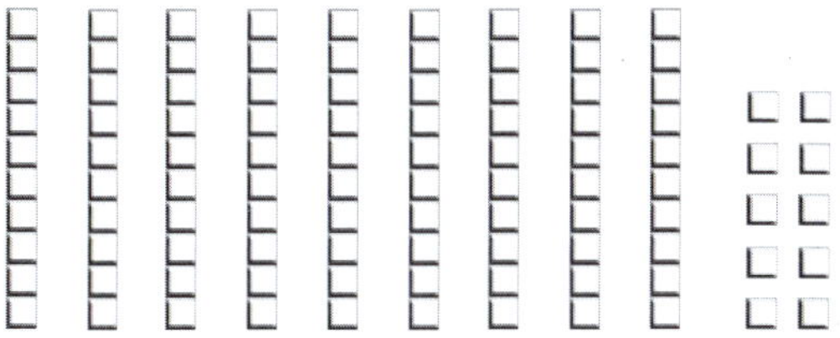

   ___ones = ___tens ___ones

4. In 1–3 ✔ the groups with more.

Count by 1s. Write the missing numbers.

5. 54, 55, 56, _57_, ___, 59, 60, ___, ___

6. 91, 92, ___, ___, 95, ___, 97, 98, ___

7. 67, ___, 69, ___, 71, ___, ___, 74, ___

Name _____

Hundred Chart Patterns

Lee can model a **number pattern**.

1	2	3	4	5	6	7	8	9	10
11	12	13	14	15	16	17	18	19	20
21	22	23	24	25	26	27	28	29	30
31	32	33	34	35	36	37	38	39	40
41	42	43	44	45	46	47	48	49	50

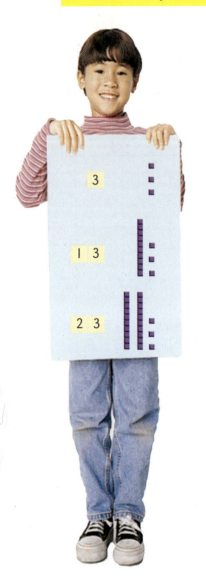

 Describe the number pattern of each color above.

Write the missing numbers.
Describe the green and orange patterns.

51	52	53	54	55					60
61		63				67			
71		73		75		77		79	
	82	83	84		86		88		
91		93						98	99	100

Find the pattern. Write the missing numbers.

1. 89, 90, 91, ___, ___
2. 62, 64, 66, ___, ___
3. 6, 16, 26, ___, ___
4. 79, 78, 77, ___, ___
5. 85, 75, 65, ___, ___
6. 98, 88, 78, ___, ___

 Use ▬▬▬ and ■ to model a pattern.

Have your child find two patterns in the hundred chart and explain how they were made.

two hundred eleven **211**

Complete each ▢. Hint: Think of a hundred chart.

50		*53*	
	61		
		72	
80			

33		
43	44	
53	54	55

64	65	66
74		
		86
	95	

Connect the dots in order. Begin at 0.

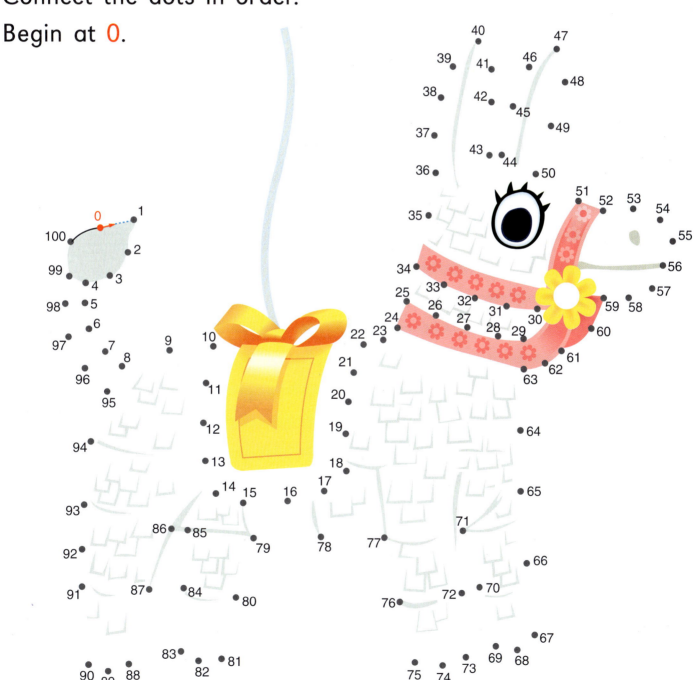

Name _____

Count On: Number After

You can name the number just after by counting on by 1s.

37, 38, 39, 40

← 37 38 39 40 →

40 is just after 39.

Write the number that comes just after.

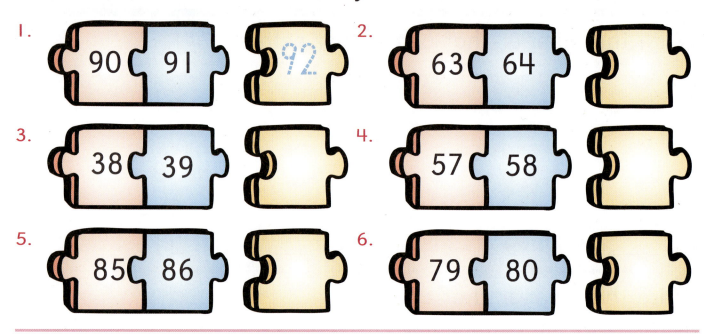

1. 90, 91, 92
2. 63, 64, ___
3. 38, 39, ___
4. 57, 58, ___
5. 85, 86, ___
6. 79, 80, ___

Write the **output** number.

7.
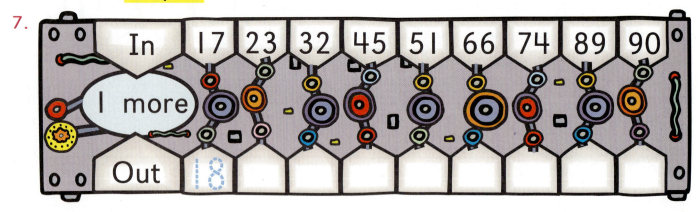

| In | 17 | 23 | 32 | 45 | 51 | 66 | 74 | 89 | 90 |

1 more

| Out | 18 | | | | | | | | |

8. I am an even number. Is the number just after me odd or even?

_____

9. I am an odd number. The number just after me is 26. What number am I?

_____

Have your child pick two 0–9 number cards, form a 2-digit number, and tell what number comes just after.

two hundred thirteen   213

Count Back: Number Before

You can name the number *just before* by *counting back*.

100, 99, 98, ...

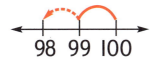

98 is just before 99.

Write the number that comes just before.

1. __73__, 74, 75
2. ____, 50, 51
3. ____, 13, 14
4. ____, 28, 29
5. ____, 91, 92
6. ____, 20, 21

Write the number that comes out of the *just before* machine.

7.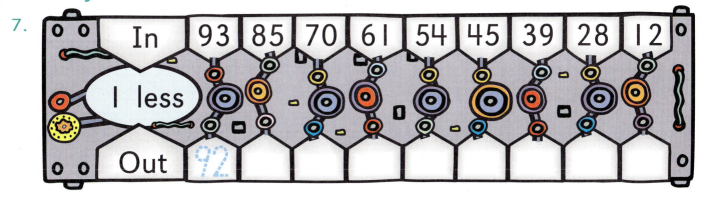

In: 93 85 70 61 54 45 39 28 12
I less
Out: 92

 Write *always* or *never*.

8. The number just before an odd number is _____ odd.

9. A number that has 5 ones is _____ just before a number with 6 ones.

Name _____

Before, Between, After

36 is just before 37.

38 is just after 37.

37 is between 36 and 38.

Write the number that comes between.

Between

1. before   after
   71, __72__, 73
   52, ____, 54
   80, ____, 82
   46, ____, 48
   83, ____, 85

Between

2. before   after
   37, ____, 39
   25, ____, 27
   64, ____, 66
   89, ____, 91
   98, ____, 100

Write the numbers that each number is between.

3. before   after
   __57__, 58, __59__
   ____, 15, ____
   ____, 89, ____
   ____, 21, ____

4. before   after
   ____, 30, ____
   ____, 62, ____
   ____, 44, ____
   ____, 76, ____

5. Is an even number always between two odd numbers?

6. Is an even number always between two even numbers?

6-13 Write 5 numbers between 0 and 100. Have your child describe each number by using the terms *before*, *between*, and *after*.

two hundred fifteen **215**

DO YOU REMEMBER?

Name _____

Subtract. Color odd differences yellow.
Color even differences purple.

Make a wheel for subtracting from 7.

Name _____

Greater Than, Less Than

Compare 54 and 36.

54 has more tens.

54 is greater than 36.

54 > 36

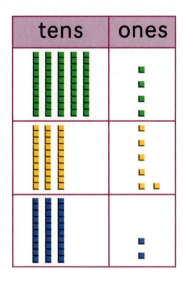

Compare 36 and 32.

Both have 3 tens.
32 has fewer ones.

32 is less than 36.

32 < 36

 Why do you compare tens first?

Write the numbers. Compare. Ring the greater.

1. | tens | ones |
    |------|------|
    | (4 green tens) | (3 green ones) |
    | (5 yellow tens) | (2 yellow ones) |

    43
    (52)

2. | tens | ones |
    |------|------|
    | (4 green tens) | (7 green ones) |
    | (3 yellow tens) | (5 yellow ones) |

    ___
    ___

3. | tens | ones |
    |------|------|
    | (5 green tens) | (3 green ones) |
    | (6 yellow tens) | (5 yellow ones) |

    ___
    ___

4. | tens | ones |
    |------|------|
    | (5 green tens) | (4 green ones) |
    | (5 yellow tens) | (1 yellow one) |

    ___
    ___

5. | tens | ones |
    |------|------|
    | (6 green tens) | (5 green ones) |
    | (4 yellow tens) | (2 yellow ones) |

    ___
    ___

6. | tens | ones |
    |------|------|
    | (6 green tens) | (3 green ones) |
    | (6 yellow tens) | (7 yellow ones) |

    ___
    ___

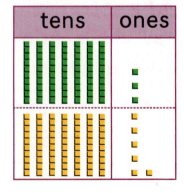 You and your child each pick 2 cards from 2 sets of 0–9 number cards and create the greatest number possible. Have your child tell which is greater.

two hundred seventeen **217**

Compare. Ring < or >. You can model each.

1. 17 (<) 27   40 < / > 30

Remember:
is less than <
is greater than >

2. 32 < / > 62   45 < / > 15   59 < / > 49

3. 26 < /(>) 25   43 < / > 41   35 < / > 38

4. 68 < / > 86   47 < / > 74   21 < / > 14

MENTAL MATH

5. Name three numbers less than 31.

6. Name three numbers greater than 60 and less than 70.

7. Name three numbers greater than 58.

Name _____

Count by Fives

I use a hundred chart to count by 5s.

1	2	3	4	5	6	7	8	9	10
11	12	13	14	15	16	17	18	19	20
21	22	23	24	25	26	27	28	29	30
31	32	33	34	35	36	37	38	39	40
41	42	43	44	45	46	47	48	49	50
51	52	53	54	55	56	57	58	59	60
61	62	63	64	65	66	67	68	69	70

1. Color every fifth number.

2. Write the numbers you colored in order.

   5, 10, ____, ____, ____, ____, ____,

   ____, ____, ____, ____, ____, ____, ____

3. Look at the ones place of the numbers in 2. What pattern do you see? _____

4. Continue to count by 5s. Predict the numbers between 70 and 100 that you would color. Write them.

   70, ____, ____, ____, ____, ____, 100

5. Write the numbers shown below in order.

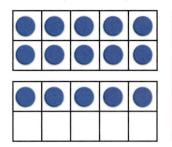 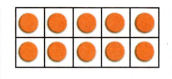 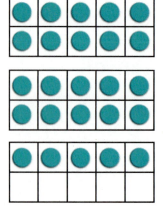 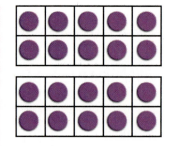

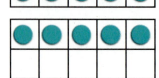

10, ____, ____, ____

Give your child 25 pennies. Have her/him make stacks of 5 pennies and use them to count by 5s.

two hundred nineteen **219**

Count by 5s. Write the missing numbers.

1.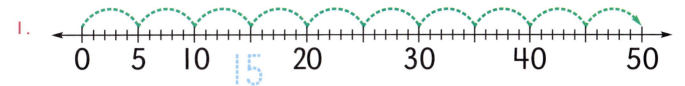
   15 ___ ___ ___ ___

2.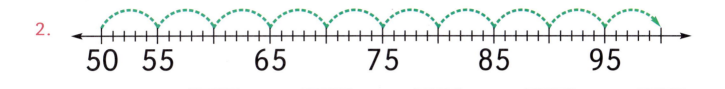
   ___ ___ ___ ___ ___ ___

3. ___, 15, ___, ___, 30, 35, ___, 45, ___

4. 60, ___, 70, ___, ___, 85, 90, ___, 100

5. 35, ___, 45, ___, ___, 60, ___, 70, ___

 PROBLEM SOLVING Draw a picture.

6. You made 4 🎀.
   You put 5 ● on each.
   You used ___ ●.

7. I have 6 🪁.
   I buy 5 🎀 for each tail.
   I buy ___ 🎀 in all.

 CHALLENGE 8. Count back. Write the missing numbers.
85, 80, 75, ___, 65, ___, ___, 50, ___, 40

220 two hundred twenty

Name _____

Count by Twos

Count by 2s.

Count by 2s.
2, 4, 6, ...

<u>2</u> <u>4</u> ___ ___ ___

___ ___ ___ ___ ___

How many in all? ___ in all

Talk It Over Describe the number pattern above.

Count by 2s. Write how many in all.

1. How many ?

   ___ ___ ___ ___ ___ ___

   ___ in all

2. How many ?

   ___ ___ ___ ___ ___ ___ ___ ___

   ___ in all

3. How many ?

   ___ ___ ___ ___ ___ ___ ___

   ___ in all

6-16 Have your child draw 8 birds. Then have her/him count by 2s and tell how many wings in all.

two hundred twenty-one **221**

Count by 2s. Write the missing numbers.

1.

1	2	3	4	5		7		9	
11		13		15	16	17		19	
21		23		25		27		29	
31		33		35		37		39	40
41		43		45		47		49	
51		53		55		57	58	59	
61		63	64	65		67		69	
71		73		75		77		79	
81		83		85		87		89	
91		93		95		97		99	100

Are the numbers in ▇ odd?
How can you tell?

Write missing numbers.
Color the odd trains green and the even ones yellow.

2. 66, 68, __, __, 72, __, __, __, __

3. 51, 53, __, __, __, __, __, __, __

4. 29, __, __, __, 35, 37, __, __, __

5. 84, __, __, __, __, 92, 94, __, __

222 two hundred twenty-two

Equal Groups

Name _____

These groups are equal.

> Equal groups have the same number in each group with none left over.

3 groups of 2 🟡 = 6 🟡

3 twos = 6

Make equal groups. Write the missing numbers.

1.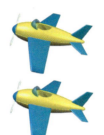

   ____ groups of __2__ = ____

   ____ twos = ____

2.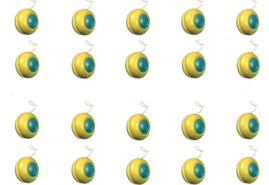

   ____ groups of __10__ = ____

   ____ tens = ____

3.

   ____ groups of __5__ = ____

   ____ fives = ____

TALK IT OVER What other ways can you make equal groups for 1–3?

6-17 Give your child 20 countables. Have him/her show at least two different ways to make equal groups.

two hundred twenty-three **223**

Jamal has 8 .
He can put them in
equal groups of 2.

4 groups of 2 = 8

4 twos = 8

Write Yes or No. Model with 🟩 or 🪙.

1. 18 🔵

   Can you make
   equal groups of 2? _____

   ___ groups of ___ = 18

   ___ twos = ___

2. 30 🪙

   Can you make
   equal groups of 5? _____

   ___ groups of ___ = ___

   ___ fives = ___

   Can you make
   equal groups of 10? _____

   ___ groups of ___ = ___

   ___ tens = ___

3. 15 🎈

   Can you make
   equal groups of 5? _____

   ___ groups of ___ = ___

   ___ fives = ___

4. Draw 12 in groups of 3.

   ___ groups of ___ 🙂

   ___ threes = ___

5. Draw 12 🙂 in groups of 4.

   ___ groups of ___ 🙂

   ___ fours = ___

Name _____

Sharing

Three friends share 15 .
Each gets an equal number.
How many does each get?

> Sharing means finding how many in each group.

Tally to find the number in each group.

Each gets __5__ .

Share. Tally. Write how many are in each group.

1. Five share 10 🖍green.

   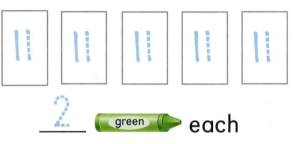

   __2__ green each

2. Four share 20 🖍orange.

   ____ orange each

3. Two share 18 🖍purple.

   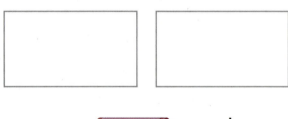

   ____ purple each

4. Two share 12 🖍yellow.

   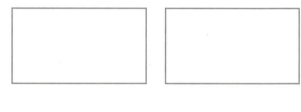

   ____ yellow each

5. Three share 12 🖍red.

   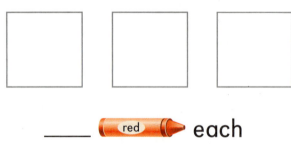

   ____ red each

6. Four share 16 🖍blue.

   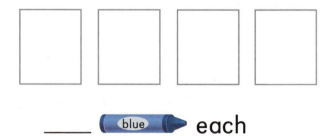

   ____ blue each

Talk It Over: How can you use 🟩 to act out sharing?

 6-18 Have your child use countables to act out how 3 share 15.

two hundred twenty-five **225**

Share. Draw an equal number in each.

1. 8 🪙 for 4 👛.
   How many 🪙 in each?

   _____ 🪙 in each.

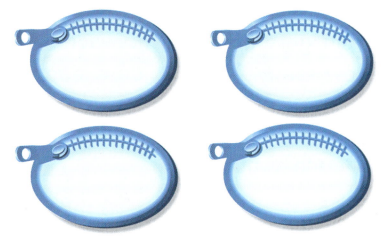

2. You have 6 🪙.
   How many can you put in each 👛?

   _____ 🪙 in each.

3. Liz puts 12 🪙 equally into 4 👛. How many are in each?

   _____ 🪙 in each.

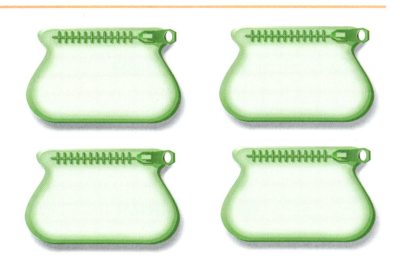

4. Toni has 8 purple. Josh has 2 purple. They share them equally. How many does Toni have then?

   _____

5. I have 9 🪙. You have 3 🪙. We share them equally. How many do you have then?

   _____

Name _____

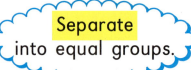

Gina makes 10 puppets.
She puts them into equal groups of 2.
How many groups can she make?

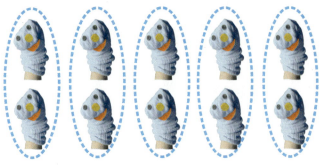

Separate into equal groups.

10 is 5 groups of two.

Ring to show equal groups.
Write how many groups.

1. 8 marbles in all
   2 in each group

   _8_ = ___ groups of two

2. 10 marbles in all
   5 in each group

   ___ = ___ groups of five

3. 12 marbles in all
   2 in each group

   ___ = ___ groups of two

4. 20 marbles in all
   10 in each group

   ___ = ___ groups of ten

6-19 Using 30 pennies, have your child make equal piles of 2s, 5s, or 10s. Ask her/him how many equal piles she/he made.

two hundred twenty-seven 227

I trade tens for ones to make equal groups.

You have 1 | 4 ▪.
Put 2 in each group.
How many groups?

14 is 7 groups of 2.

Use ▭ and ▫. Write how many groups.

1. 1 ▭ 5 ▫
   3 in each group
   _15_ = ___ groups of three

2. 2 ▭ 0 ▫
   4 in each group
   ___ = ___ groups of four

3. 1 ▭ 6 ▫
   2 in each group
   ___ = ___ groups of two

4. 2 ▭ 5 ▫
   5 in each group
   ___ = ___ groups of five

5. 2 ▭ 0 ▫
   5 in each group
   ___ = ___ groups of five

6. 1 ▭ 8 ▫
   3 in each group
   ___ = ___ groups of three

 Use ▭ and ▫.

7. Pablo buys 16 wheels. He needs 4 for each 🛒. How many 🛒 can he make?
   ____

8. Betty buys 18 beads. She puts 6 beads on each ⚬. How many ⚬ can she make?
   ____

Name _____

Calculator: Skip Counting

To count on by 10s from 40, press these keys.

I can skip count on a calculator.

Remember to clear the calculator.

| 0. | 4. | 40. | 40. | 1. | 10. | 50. | 60. | 70. | 80. |

Press these keys.
Write the numbers.

1.

   0   1   15   15   5   20   ___   ___   ___   ___

2.

   0   6   60   60   1   10   70   80   ___   ___

Talk It Over: What keys do you press to count by 2s from 30?

3. Write the keys to count by 2s from 32 to 40.

   0   3   32   32   2   34   36   38   40

This lesson teaches how to use a calculator to skip count.

To count back by 10s from 70, press these keys.

Press these keys. Write the numbers.

1. ON/AC 2 5 − 5 = = = =

   0   2   25   25   5   __   __   __   __

2. ON/AC 1 8 − 2 = = = =

   0   1   18   18   2   __   __   __   __

3. ON/AC 3 2 + 1 0 = = = =

   0   3   __   __   __   __   __   __   __   __

4. Draw the keys you press to count on by 2s from 50. Stop at 60.

5. Draw the keys you press to count back by 5s from 90 to 70.

230 two hundred thirty

PROBLEM-SOLVING STRATEGY
Guess and Test

Use these numbers and guess and test to solve each problem.

1. **Read** I am an even number greater than 10 but less than 20.

   Think Test each even number. Is it greater than 10? Is it less than 20?

   Write Which numbers in the 🎈 are even?
   10, 14, 28, 60, 66, 80

   Which of these are greater than 10?
   14, 28, 60, 66, 80

   Check Which is less than 20? 14

2. **Read** I am a number you say when you count by 5s. I am an odd number between 60 and 80.

   Think Count by 5s. Start with 10.

   Write Which numbers in the 🎈 do you say?
   10, 35, ___, ___, ___, ___

   Which of these numbers are odd?
   ___, ___, ___

   Check Which of these is between 60 and 80? ___

Play a guessing game with your child using problems like the ones above.

Read → Think → Write → Check

Solve by guess and test.

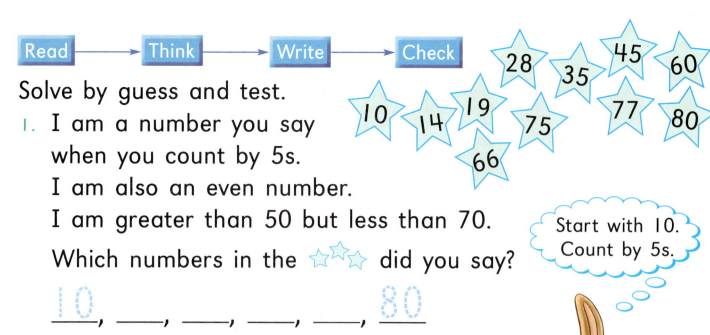

1. I am a number you say when you count by 5s.
   I am also an even number.
   I am greater than 50 but less than 70.
   Which numbers in the ☆ did you say?

   Start with 10. Count by 5s.

   10, ___, ___, ___, ___, 80

   Ring the even numbers.

   Which is greater than 50 but less than 70? I am ___.

2. I am a number you say when you count by 5s. I am an even number. I am less than 40. I am ___.

3. I am an odd number. I have the same number of ones and tens. I am ___.

4. I am a number you say when you count by 2s. I am between 20 and 30. I am ___.

5. I am a number you say when you count by 10s. I am greater than 60. I am ___.

6. I am a number you say when you count by ___.
   I am between ___ and ___.

Name _____

PROBLEM-SOLVING APPLICATIONS
Logical Reasoning

1. **Read** Milo put the numbers 71 through 80 in a box. Carol took out 72, 73, 77, 78, and 79. What numbers are still in the box?

   Think List the numbers Milo put in. ✗ the numbers Carol took out.

   71, ~~72~~, ~~73~~, 74, 75, 76, ~~77~~, ~~78~~, ~~79~~, 80

   Write Write the numbers still in the box.

   Check 71, ___, ___, ___, ___

2. Sara counts from 35 through 40. Then Darryl counts from 38 through 42. What numbers do both children count?

   List the numbers each says.

   Sara: 35, 36, ___, ___, ___, ___

   Darryl: 38, ___, ___, ___, ___

   Write the numbers both children counted. ___, ___, ___

   (35, 36, ...)
   (38, 39, ...)

3. Niah put the numbers between 85 and 93 in a box. Brian took out the even numbers. What numbers are still in the box?

   ___, ___, ___ are still in the box.

   In this lesson your child solved problems by using the *Logical Reasoning* strategy.

   two hundred thirty-three 233

Use a strategy you have learned.

STRATEGY FILE
Guess and Test
Logical Reasoning
Find a Pattern

1. Rhea wrote all the 2-digit numbers that have 7 ones. How many numbers did she write?

   17 27 37 _____

   Rhea wrote ____ numbers.

2. Joel has 20 wheels. Each robot he makes needs 4 wheels. How many robots can Joel make?

Numbers of Wheels	4				
Numbers of Robots	1				

   Joel can make ____ robots.

3. I am an odd number. The digit in my tens place is one more than the digit in my ones place. Which number am I? ____

4. Jenna's birthday is in January. It is a number you say when you count by tens. It is greater than 25. When is Jenna's birthday?

   January

			1	2	3	4
5	6	7	8	9	10	11
12	13	14	15	16	17	18
19	20	21	22	23	24	25
26	27	28	29	30	31	

   Jenna's birthday is on January ____.

PROBLEM SOLVING

Name _____

Chapter Review and Practice

Write how many.

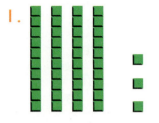

1. __4__ tens __3__ ones

   __43__

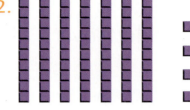

2. ____ tens ____ ones

   ____

3. ____ tens ____ ones

   ____

Write the place value.

4. 65 = ____ tens ____ ones

5. 18 = ____ ten ____ ones

6. 96 = ____ tens ____ ones

Write the number.

7. 8 tens 9 ones = ____

8. 5 tens 0 ones = ____

9. 4 tens 6 ones = ____

10. Which number is greater?

    | 13 | (15) | | 78 | 69 |

11. Which number is less?

    | 38 | 48 | | 73 | 37 |

Write the missing number.

Before

12. ____, 95

13. ____, 30

After

14. 43, ____

15. 71, ____

Between

16. 87, ____, 89

17. 39, ____, 41

Write the missing numbers.

18.

| 30 | 32 | | | 38 | | 42 | | | 48 |

19.

| 55 | 60 | 65 | 70 | 75 | | | | | 100 |

This page reviews the mathematical content presented in Chapter 6.

two hundred thirty-five 235

Name _____

Use 🟪. Make equal groups.
How many 🟪 left over?
Color.
0 left over blue.
1 left over green.
2 left over yellow.

Make equal groups. Count the cubes left over.

17 🟪 in 5 groups __2__

9 🟪 in 2 groups ____

22 🟪 in 10 groups ____

5 🟪 in 2 groups ____

31 🟪 in 10 groups ____

12 🟪 in 5 groups ____

16 🟪 in 5 groups ____

20 🟪 in 2 groups ____

30 🟪 in 10 groups ____

35 🟪 in 5 groups ____

15 🟪 in 5 groups ____

14 🟪 in 2 groups ____

6 🟪 in 5 groups ____

11 🟪 in 5 groups ____

21 🟪 in 10 groups ____

13 🟪 in 2 groups ____

Performance Assessment

Model each number. Complete each ___.

1. Thirty-one has ___ tens ___ one.

   It is just after ___ and

   just before ___.

   It is greater than ___.

2. Thirteen has ___ ten ___ ones.

   It is between ___ and ___.

   Is it odd or even? ___ Why?

3. Draw or model each group of numbers. Which group makes a pattern?

   45, 40, 38, 37   32, 22, 12, 2

Choose 1 of these projects. Use a separate sheet of paper.

4. Lee made 10 of these with these blocks:

   , ◆,

   and ■.

   How many ◆ did she use?
   How many ■ did she use?
   Explain your thinking.

5. Which group of 🪙 can you stack in twos, fives, and tens?
   Write Yes or No in the table.

pennies	twos	fives	tens
12			
20			
15			

two hundred thirty-seven **237**

Check Your Mastery

Name _____

Write the place value. Then write the number.

1.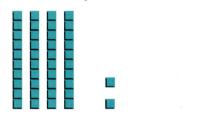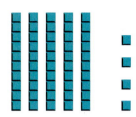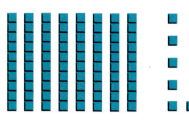

   ___tens ___ones   ___tens ___ones   ___tens ___ones

   _____   _____   _____

2. forty ___tens ___ones   ninety-six ___tens ___ones

   _____   _____

Write the number.

3. 3 tens 8 ones _____   8 tens 1 one _____

Write the missing numbers.

4. ___, 50   5. 36, ___   6. 13, ___, 15

Ring the even numbers. | Ring the odd numbers.

7. 21 22 23 24   8. 17 22 36 41

Count by 5s. Write the numbers.

9. 5, 10, ___, 20, ___, 30, 35, ___, 45, ___

10. In 9 ring the count by 10 numbers.

 11. I am between 40 and 50.
    I am a number you say
    when you count by 2.
    I have less than 3 ones.   I am ___.

Money and Time

7

CRITICAL THINKING

Today is Friday. The day before yesterday Hanna bought the sale item. What item did she buy? What day did she shop?

For more information about Chapter 7, visit the Family Information Center at **www.sadlier-oxford.com**

Dear Family,

Today your child began Chapter 7. As he/she studies money and time, you may want to read the poem below, which was read in class. Encourage your child to talk about some of the math ideas shown on page 239.

Look for the 🏠 at the bottom of each skills lesson. The suggestion on the page gives you an opportunity to improve your child's understanding of math. You may want to have coins and clocks available for your child to use throughout this chapter.

Home Reading Connection

Spending Time and Pennies

Mom gave me a nickel
Dad gave me a dime.
I found a shiny quarter
And now I'm feeling fine.

Gram gave me a wristwatch
That tells the time with light.
Pop gave me a ticking clock,
Its hands move day and night.

I'll count the pennies in my bank
I'll count them two by two.
Then I'll take my pennies
And give half to you.

At 12 o'clock tomorrow
I know just what we'll do.
We'll spend our time and pennies
And shop for something new!

Helen T. Smythe

Home Activity

Time Adds Up

Try this activity with your child. Make a clock face on a paper plate. Have your child place 5 pennies next to each hour and count by 5s to 30. Then have her/him trade pennies for nickels and nickels for dimes as these concepts are introduced in the chapter.

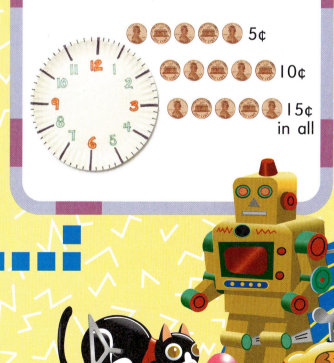

Name _____

MATH CONNECTIONS
Graphs/Visual Reasoning

Count the coins in each purse.
Make a graph for each. Color 1 □ for each coin.

Number of Coins

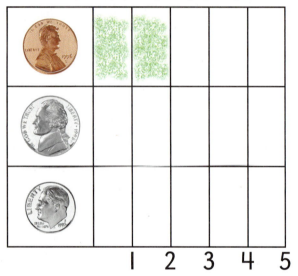

Number of Coins

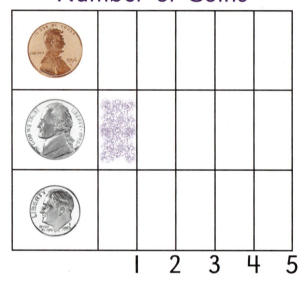

Which purse can you take a coin from to get each?
Color the purse.

1. always a nickel
2. sometimes a dime
3. never a nickel
4. sometimes a penny

5.

6.

You can put this page in your Math Portfolio.   two hundred forty-one **241**

CROSS-CURRICULAR CONNECTIONS
Language Arts

Name _____

Use **ordinals** to put each sentence in order.

1. [watch] [My] [is] [red.]
   2nd   1st   3rd   4th

2. [two.] [clock] [The] [struck]
   ___ ___ 1st ___

3. [time] [it?] [What] [is]
   ___ ___ 1st ___

4. [costs] [ticket] [20¢.] [One]
   ___ ___ ___ 1st

5. [dime] [A] [is] [small.]
   ___ 1st ___ ___

6. [Watches] [two] [hands.] [have]
   1st ___ ___ ___

> The **first** word in a sentence has a capital letter. A period or question mark is at the end.

> You can write ordinals in two ways.

Ordinals
first — 1st
second — 2nd
third — 3rd
fourth — 4th

Scramble a sentence. Then put the words in order.

7. [ ] [ ] [ ] [ ]

___ ___ ___ ___ ___

242 two hundred forty-two

You can put this page in your Math Portfolio.

Name _____

Count on by Pennies

You have pennies and nickels.
First count the nickels by 5s.
Then count on by 1s for the pennies.

2 nickels and 3 pennies equals 13¢.

Say   5¢,   10¢,   11¢,   12¢,   13¢   **amount**

Write how much.

1.   9¢

2.   ____ ¢

3.   ____ ¢

4.   ____ ¢

 In 1–4 ✔ the least amount.

Ring the amount.

5. 18¢

 Why do you sort coins before counting?
Which would you rather have: 2 pennies 4 nickels or 4 pennies 2 nickels?

Give your child a mixed group of nickels and pennies or dimes and pennies to sort and count.

First I count the dimes by 10s.
Then I count on by 1s for pennies.

Say 10¢, 20¢ 30¢, 31¢, 32¢

Write how much.

1.   15¢

2.   _____ ¢

3.   _____ ¢

4.   _____ ¢

 ✓ the greatest amount in 1–4.

 Use , , and .

5. You can buy with 4 coins. One coin is a penny. The other 3 coins are _____.

6. Lily has 3 nickels. How many pennies does she need to buy each?

    _____   _____

Name _____

Quarters

1 quarter = 25 cents

25¢

I have 25¢ in each hand.

Count on by a penny. Write the amount.

1.

   25¢, 26¢, 27¢   27¢

2.   ____¢

3.   ____¢

4.   ____¢

5.   ____¢

 TALK IT OVER   Which shows more than 30¢?

Provide your child with a quarter and some pennies and ask her/him to show you a specific amount of money.

7-4   two hundred forty-seven

Ring the amount you spend.
Write the amount you have left.

Amount Left

1. 27¢ — (quarter + penny + penny) ringed; penny, dime remain — 11¢

2. 28¢ — quarter + penny + penny + penny + nickel — ___ ¢

3. 32¢ — quarter + 10 pennies — ___ ¢

4. 29¢ — quarter + 5 pennies — ___ ¢

5. Draw coins to show 4 ways to make 25¢.

Name _____

Count on by Nickels

Ramon sorts dimes and nickels.
He counts on by 10s and 5s.

Say 10¢,   20¢,   30¢,   35¢,   40¢

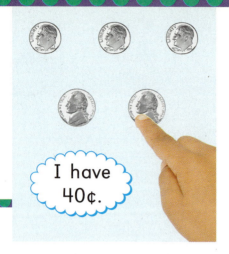

I have 40¢.

Write how much.

1.   _35¢_

2.   ____¢

3.   ____¢

4.   ____¢

5.   ____¢

6.   ____¢

7.   ____¢

 TALK IT OVER You can pick 4 coins.
Which would you pick? Why?

 Listen as your child counts on aloud some of the exercises in this lesson.

two hundred forty-nine

Count on by a nickel.

1.

   25¢, 30¢, _____¢, _____¢, _____¢

2.   _____¢

3.   _____¢

Ring the amount.

4. 40¢

5. 35¢

6. 45¢

 CHALLENGE Write more or less.

7.

   _____ than 30¢

8.

   _____ than 50¢

250 two hundred-fifty

Name _____

Count Mixed Coins

Yuri has 1 , 1 ,
1 , and 2 .

Put the coins in **order**.
Then count on by 10s, 5s, and 1s.

Say   25¢,   35¢,   45¢,   50¢,   51¢

I have 51¢.

Write how much.

1.    16¢

2.    ____ ¢

3.    ____ ¢

4.
   ____ ¢

5.
   ____ ¢

6.
   ____ ¢

SECOND LOOK In 1–6 ✓ amounts between 35¢ and 45¢.

MATH JOURNAL 7. Draw 3 dimes, 1 quarter, 2 pennies, and 1 nickel in the easiest order to count on.

7-6 Ask your child to pick out and count 4 coins from a bag containing 1 quarter, 4 dimes, 4 nickels, and 4 pennies.

two hundred fifty-one **251**

Use the table and coins.
How much money does each child have?

Coins Saved

	quarter	dime	nickel	penny	Total
Kimi	1	2	1	5	¢
Zoe	0	4	4	0	¢
Pat	1	1	2	5	¢
T.J.	1	1	4	0	¢

1. Who has the most money? _____

2. Who has the least money? _____

3. Which two have the same amount? _____
   _____

4. Kimi and Pat both have ___ coins. Who has more money? _____

 Do not count.
✔ the amount closest to 50¢.

5.

6.

Trading

You can **trade** equal amounts.

2 nickels = 10¢

1 dime = 10¢

1 nickel 5 pennies = 10¢

 Tell another fair trade for 10¢.

1st Write how much.

2nd Match equal amounts.

1.  _15_¢

2.  ____¢

3.  ____¢

4.  _15_¢

5. ____¢

6.  ____¢

7.  ____¢

8.  ____¢

9.  ____¢

10.  ____¢

Tell your child to show 17¢ in three different ways.

✔ the fair trade. Tell why the other is not fair.

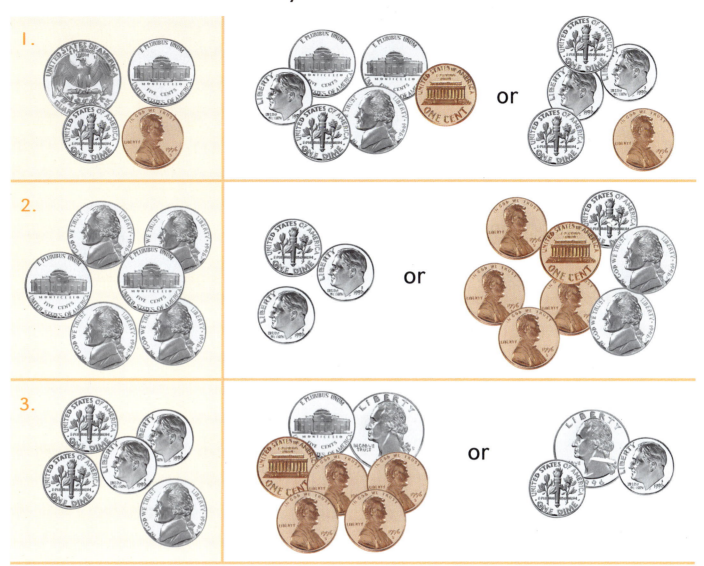

 Draw the coins.

4. Leah has 3 dimes and 4 pennies. How can Leah trade to share the money equally with Jay?

Leah | Jay

5. Ian had 3 nickels and 1 penny. Yona had 4 pennies. They traded to share. Which coins does each have now?

Ian | Yona

Name

Using Money

Tad has 2 dimes, 1 nickel, and 2 pennies. What can he buy?

10¢, 20¢, 25¢, 26¢, 27¢

Tad can buy a .

 Name different coins you could use to buy 🔷.
Name 2 ways you can pay for 💎.

Write the amount. Match.

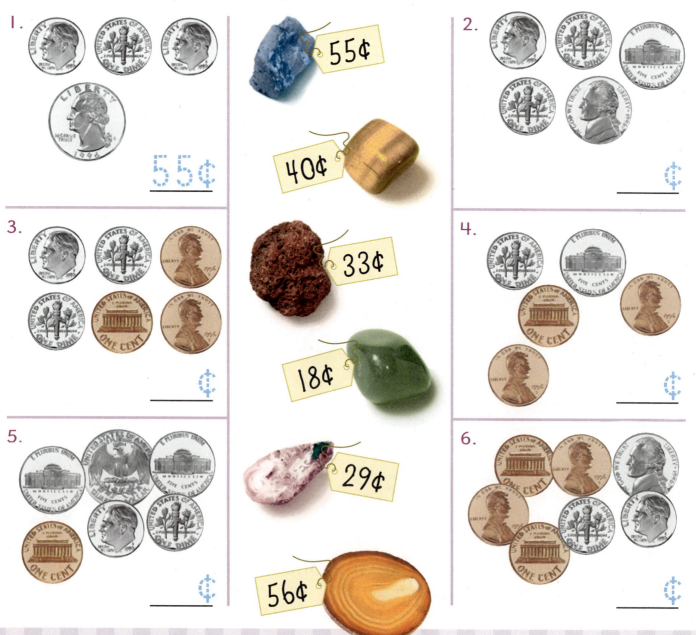

Put price tags under 69¢ on various items. Have your child name the coins he/she needs to buy each item.

two hundred fifty-five 255

Write the amount you have to spend.

Draw ☺ if you have enough money.
✗ coins you do not need.

Draw ☹ if you need more money.
Color coins you need.

I need 1¢ more.

6. Make a chart like the one above to buy 3 things.

Name _____

Equal Amounts

Do Lulu and Jon have **equal amounts**?

Trade and compare to make the amounts equal.

Now Lulu has 2 dimes 1 nickel 1 penny.
Now Jon has 2 dimes 1 nickel 0 pennies.
Jon needs 1 🪙 for an equal amount.

5 pennies = 1 nickel
2 nickels = 1 dime

Trade. Complete the table.

1. Sara   ___  ___

   Scott  ___  ___

   Scott needs ___ _____.

2. Lee  ___  ___ ___

   Yuka  ___  ___  ___

   Lee needs ___ _____.

3. Lisa    ___

   Juan  ___  ___

   Lisa needs ___ _____.

Name _____

Write Y for yes.
Write N for no.

Color:
 Y
 N

Does 60 come next in 100, 90, 80, ☐? N

Does 31 equal thirteen?

Is 31 just after 30?

Do 8 tens 5 ones = 85?

Does 1 less than 99 = 89?

Is 83 less than 87?

Is 28 ten more than eighteen?

Are 66, 68, and 70 in order counting by 2s?

Is 45 greater than 55?

Is 18 between 17 and 19?

Does 2 tens = thirty?

Does 10 less than 45 = 35?

Does 18 = 7 twos?

Does thirty = 50?

Do 5 groups of 10 = 50?

Do 4 groups of 5 = 20?

Is 21 just before 20?

Are 81 and 83 the missing numbers in ☐, 82, ☐?

Does 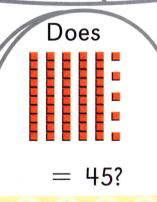 = 45?

Are there 0 left over when 2 share 10 pennies equally?

258 two hundred fifty-eight

Hour

The long minute hand points to the 12.

The short hour hand points to the 7.

7 o'clock

7 o'clock

The : is between the hours and the minutes.

 How are both clocks the same? How are they different?

Write the time in 2 ways.

1.
2.
3.
4.

__12__ o'clock   ____ o'clock   ____ o'clock   _____

`12:00`   `:00`   `:`   `:`

5. (purple clock)   6.    7. 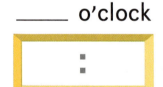   8.

____ o'clock   ____ o'clock   ____ o'clock   _____

`:`   `:`   `:`   `:`

 Name 1 thing you do at the times in 1–8.

Show the time. Draw the hour hand.

1.  2 o'clock

2.

3.  7 o'clock

4.

5.  8 o'clock

6.

7.  11 o'clock

8.

9. The town clock chimes 2 times at 2 o'clock and 3 times at 3:00. How many times will it chime at 4 o'clock?

It will chime ____ times.

10. Write the missing numbers.
11. Draw 6 o'clock.

Name _____

Half Hour

Both clocks show two thirty.

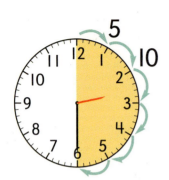

The hour hand is halfway between 2 and 3.

The minute hand is halfway around the clock.

2:30
half past 2
30 minutes after 2

Write the time in 2 ways.

1.

half past __10__

2.

____ o'clock

3.

____ minutes after ____

4.

half past ____

5.

____ thirty

6.

____ minutes after ____

 Which times in 1–6 are in order?
How many minutes are in 1 hour?

 Listen as your child describes where the hour and minute hands are for three thirty and five thirty, names the times, then draws them.

two hundred sixty-one **261**

Draw the missing hands. Write the time.

1.  30 minutes after 3

2.  half past 11

3.  half past 8

4.  9 o'clock

5.  30 minutes after 10

6.  one thirty

7.  half past 12

8.  30 minutes after 7

 CRITICAL THINKING

Draw the time.
Write the order: 1st, 2nd, 3rd.

9.
half past 2

11.
one thirty

2 o'clock

___ ___ ___

262 two hundred sixty-two

Name _____

Time Patterns

These clocks show time patterns.

Draw the time that comes next.

1.

   half past ____   ____ o'clock   ____ minutes after ____

2.

   ____ o'clock   ____ o'clock   ____ o'clock

3.

   ____ o'clock   half past ____   ____ o'clock

4.

   half past ____   ____ o'clock   ____ minutes after ____

 ✓ the half-hour patterns 1–4.

Say a half-hour time sequence and ask your child to say and draw what comes next.

two hundred sixty-three **263**

Write the time. Color the matching ==digital clock==.

1.  half past __12__ or

   ___ minutes after ___

2.  ___ o'clock

3.  ___ minutes after ___

   or half past ___

4.  half past ___ or

   ___ minutes after ___

5.  ___ minutes after ___

   or half past ___

 Use 🕐 and coins.

6. You win one penny every minute. You win for 1 half hour. Write how you can trade your pennies.

   ___ 🪙 or ___

   or ___

264 two hundred sixty-four

Name _____

Name the Time

It is 7:30.
I leave at 8:00.

School starts at 8:30.

30 minutes is 1 half hour.   60 minutes is 1 hour.

7:30 to 8:30 is 1 hour.

Draw the time.

1. Practice begins at 6 o'clock.   Practice ends at 7 o'clock.

_____6:00_____ to _____7:00_____ is ____ hour.

2. We start at 3:30.   We finish at 4:30.

_____ to _____ is ____ hour.

Help your child set an alarm clock to ring 1 hour after 7 o'clock or 1 hour after 6:30.

two hundred sixty-five **265**

Cut and paste to match.
Write the order: 1st, 2nd, 3rd, or 4th.

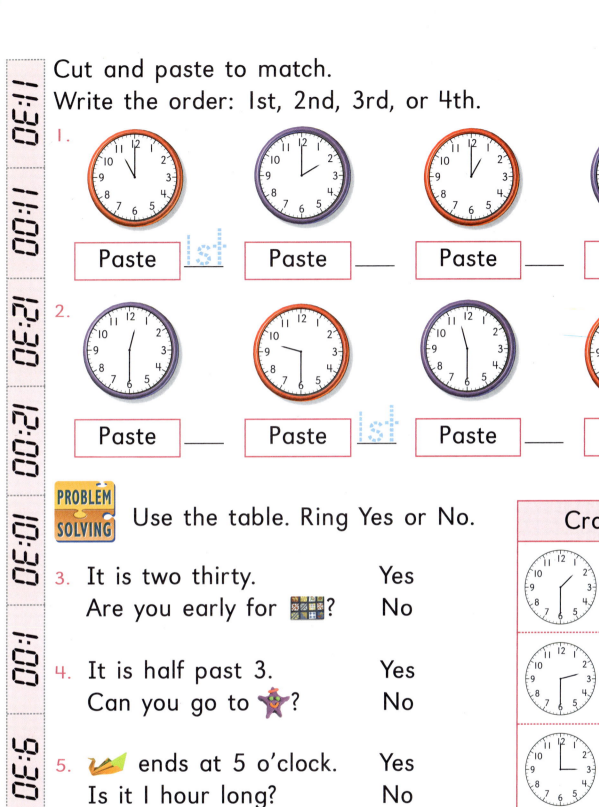

Problem Solving Use the table. Ring Yes or No.

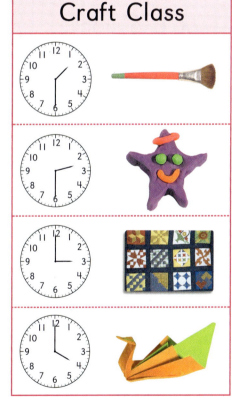

Craft Class

3. It is two thirty.   Yes
   Are you early for 🧩?   No

4. It is half past 3.   Yes
   Can you go to ⭐?   No

5. 🦢 ends at 5 o'clock.   Yes
   Is it 1 hour long?   No

6. 🖌 is 1 hour long.   Yes
   Can you go to both   No
   🖌 and ⭐?

7. Write a schedule. Tell whether each activity is 1 hour long.

266 two hundred sixty-six

Name _____

Estimate Time

"I can write A to Z in **about 1 minute**."

"It takes **about 1 hour** for me to draw a picture for each of the letters."

Color about how long.

1. setting the table

6:00

[colored] about 1 minute

about 1 hour

2. going to the dentist

4:00

about 1 minute

about 1 hour

3. baking a chicken

4:30

about 1 minute

about 1 hour

4. brushing your teeth

7:30

about 1 minute

about 1 hour

Which activites in 1–4 go together?
About what time will the activity end in 2 and 3?

7-14 Tell your child to draw a picture of some activities that take about 1 minute and others that take about one hour.

two hundred sixty-seven **267**

Color the box.

1. making the bed

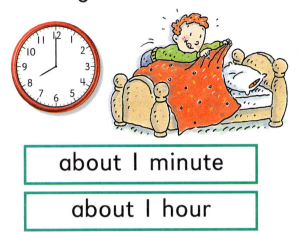

   about 1 minute

   about 1 hour

2. playing a game

   about 1 minute

   about 1 hour

3. shopping for food

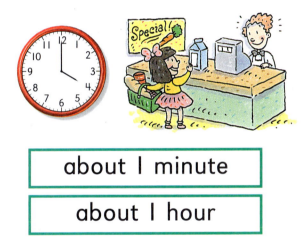

   about 1 minute

   about 1 hour

4. tying your shoes

   about 1 minute

   about 1 hour

5. Soccer practice is about 1 hour. It begins at 6:30. When will practice end?

   _____

6. Téa irons for 1 hour. She irons 3 days a week. How long does she iron this week?

   _____

7. Tara comes to school first. Tim comes 1 half hour after Ed. Ring the clock to show the time Ed gets there.

Name _____

Calendar

A calendar shows the days in each month.

 How many months are in 1 year?
Name the 2nd, 3rd, and 4th months.

❄❄❄❄❄❄❄ January ❄❄❄❄❄❄❄						
Sunday	Monday	Tuesday	Wednesday	Thursday	Friday	Saturday
	1 HOLIDAY	2	3	4	5	?
7	?	9	?	11	?	13
14	15	16	17	18	19	20
?	22 HOLIDAY	23	24	25	26	27
28	?	?	31			

Name the missing dates.

Color

1. weeks with 7 days [blue].

2. first and last day of month [yellow].

Complete.

3. January has __31__ days and ____ full weeks.

4. It begins on _____, January ____.

5. It has ____ Mondays and ____ Saturdays.

6. The third Friday is January ____.

7. There are ____ holidays and ____ school days.

8. February begins after January ____.

 Observe as your child finds tomorrow's date on a calendar and writes the day and date.

FINDING TOGETHER Use a blank calendar.
Make a calendar for February.

> February has 28 days.

1. February 1 is the day after January 31.

2. _____ is February 1.

3. The month ends on _____, February 28.

Write how many.

4. Thursdays ____   5. Sundays ____   6. full weeks ____

7. Color February 14 red.

Write the date.

8. 2nd Monday _____.

9. 3rd Saturday _____.

PROBLEM SOLVING Use the calendar you made.

10. Rob has a music lesson every Tuesday. Rob has ____ music lessons in February.

11. Snow begins on the third and ends on the tenth of February. It snows ____ days altogether.

Today is February 5.

12. What is the day after tomorrow?

_____

13. What was the day before yesterday?

_____

270 two hundred seventy

Name _____

PROBLEM-SOLVING STRATEGY
Logical Reasoning

1. Starting time
9 o'clock

Ending time

8:00

3:30

2. Breakfast
half past 7

Finished at

8:00

9:00

3. Bedtime
8:00

Wake up at

7:00

4:00

4. Starting time
2 o'clock

Time to go

10:30

3:30

Say a starting time for a familiar activity. Have your child give a reasonable ending time of from 1 to 3 hours later.

two hundred seventy-one **271**

Read → Think → Ring

1. Arrive
   half past 11

   Home

2. Starting time
   1 o'clock

   Finished at

3. Enter store
   half past 9

   Leave store

4. Starting time
   30 minutes after 12

   Finished at

PROBLEM SOLVING

PROBLEM-SOLVING APPLICATIONS
Hidden Information

1. **Read** I have a dozen pennies.
   I lose 8 of them.
   How much money is left?

   Think __12__ pennies in a dozen

   Write __12¢__ ◯ __8¢__ = ____ ¢

   Check I have ____ ¢ left.

   I dozen = 12 things. The 1st sentence has hidden information.

2. **Read** Rita has 4¢. She finds a nickel and a penny. How much money has she in all?

   Think A nickel and a penny equals ____.

   Write → **Check** Rita has ____ in all.

   4¢
   + ___

3. I have 1 quarter 2 dimes.
   I trade the coins for nickels.
   How many nickels do I have?

   1 quarter = ___ nickels, 1 dime = ___ nickels

   I have ___ nickels.

4. Milo worked a dozen hours.
   He worked 4 hours on Monday and the rest on Tuesday. How many hours did he work on Tuesday?

   1 dozen = ___.

   Milo worked ___ hours on Tuesday.

Use a strategy you have learned.

STRATEGY FILE
Find a Pattern
Hidden Information
Logical Reasoning
Choose the Operation

1. Our trip started Monday, June 4. It ended 2 weeks later. When did our trip end?

   Our trip ended on *Monday, June* _____.

2. On Monday I saved 2 pennies. Every day since then I saved 2 pennies. How much will I have altogether by Friday?

   On Friday I will have _____.

3. Carmen has 12¢.
   She spends a nickel.
   How much money is left?

   Carmen has _____ left.

4. Jill gets home last. Jack comes in 1 half hour after Todd. Todd gets home first. What time does each come home?

   Jill ___:___   Jack ___:___   Todd ___:___

5. Kim has 1 quarter. He finds 1 nickel. How much does he have now?

   Kim has _____ now.

6. 2 of me equals a dime.
   5 of me equals a quarter.
   What coin am I?

   _____

Name _____

Cumulative Review II
Chapters 1–7

Mark the ○ for your answer.

Listening Section

A

2:00   7:00   11:00
 ○   ○   ○

B

○   ○   ○

---

1. How many?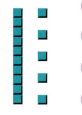
   ○ 17
   ○ 15
   ○ 13
   ○ 12

2. How many?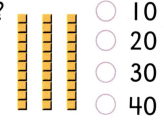
   ○ 10
   ○ 20
   ○ 30
   ○ 40

3. Count by twos.
   What comes next?

   2, 4, 6, 8, ___

   9   10   11   12
   ○   ○   ○   ○

4. What comes between?

   56, ___, 58

   66   59   57   55
   ○   ○   ○   ○

5. 5 tens 3 ones equals

   33   35   50   53
   ○   ○   ○   ○

6. Which is less than 42?

   39   43   51   60
   ○   ○   ○   ○

7. What comes next?

   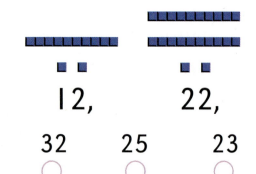

   2,   12,   22, ___

   42   32   25   23
   ○   ○   ○   ○

8. What is a fair trade for this coin?

   10¢   5¢   10¢   5¢
   ○   ○   ○   ○

REINFORCEMENT

two hundred seventy-nine   **279**

Mark the ○ for your answer.

9. What time is it?

   ○ half past 6 ○ 6 o'clock
   ○ 2 o'clock   ○ two thirty

10. What number is missing?

    $8 - 8 = \square$
    $8 - \square = 8$

    16   4   0   2
    ○   ○   ○   ○

11. $7 + 3 + 2 =$ ___

    4   9   12   11
    ○   ○   ○   ○

12. $11 - 5 =$ ___

    6   7   8   9
    ○   ○   ○   ○

13. What number is missing?

    ___, 92, 93, 94

    99   96   95   91
    ○   ○   ○   ○

14. Show sums of 9.

    $6 + 3 = 4 +$ ___

    5   6   7   8
    ○   ○   ○   ○

15. Which does not belong to this fact family?

    ○ $6 + 4 = 10$ ○ $10 - 7 = 3$
    ○ $10 - 4 = 6$ ○ $10 - 6 = 4$

16. Leave for a party at this time. Come home at __?__

    ○ 10:00   ○ 1:30
    ○ 3:00   ○ 12:00

17. Tyrell and Amy had 1 dozen cookies. Each child ate 4 cookies. Now they have ___ left.

    ○ 3 cookies   ○ 4 cookies
    ○ 6 cookies   ○ 8 cookies

18. A costs 46¢. Which coins do you still need?

    ○ 2 quarters   ○ 2 nickels
    ○ 2 dimes   ○ 2 pennies

Geometry, Fractions, and Probability

8

CRITICAL THINKING

Solid figures take up space. Plane figures are flat. Are there more plane figures or solid figures on the playground?

281

For more information about Chapter 8, visit the Family Information Center at www.sadlier-oxford.com

Dear Family,

Today your child began Chapter 8. As she/he studies geometry and fractions, you may want to read the poem below, which was read in class. Encourage your child to talk about some math ideas shown on page 281.

Look for the 🏠 at the bottom of each skills lesson. The suggestion on the page gives you an opportunity to improve your child's understanding of math. You may want to have shapes and models of halves, thirds, and fourths available for your child to use throughout this chapter.

Home Activity

Ship-Shape Shapes

Try this activity with your child. Cut different-sized geometric shapes from colored paper, and ask her/him to identify each shape. Help her/him make models for playground equipment by pasting the shapes onto a large sheet of paper. For example, various shapes could form a playhouse, ship, or spacecraft. After your child finishes each lesson in this chapter, add the shapes and/or talk about the concepts presented.

Home Reading Connection

The Playground on Euclid Street

Merry-go-rounds, carousels, a train, and a slide,
Concrete cylinders, monkey bars, a bench to one side.
Atop the jungle gym, a near-perfect sphere,
Touching the sky, you can find us here.

By the hopscotch squares are the wooden swings
Next to four wild animals on rectangular springs.
Climb a ladder and shoot down the slide,
Half afraid, but enjoying the ride.

Up and down—don't fall!—look around,
Whirling in circles on a merry-go-round.
All aboard the train for a circle tour
Of the Euclid Street playground, with fun galore.

Christine Barrett

Name _____

MATH CONNECTIONS
Inside, Outside, On

Follow directions to show the position of each.

Tally and make a bar graph to show how many children were in each position.

Position	Tally	Number of Children						
inside								
outside								
on								

     1 2 3 4 5 6 7

PORTFOLIO 8 — You can put this in your Math Portfolio.

two hundred eighty-three **283**

CROSS-CURRICULAR CONNECTIONS
Multicultural Studies

Name _____

The Olympic symbol uses 1 color from the flag of every country.

 Name the shapes in the Olympic symbol.

Amish quilts use shapes, too. These quilt shapes form patterns.

Color inside each quilt pattern.

1. Use red, white, and blue for the United States.

2. Use the colors from another country's flag.

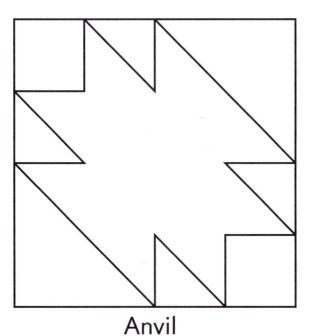

Anvil

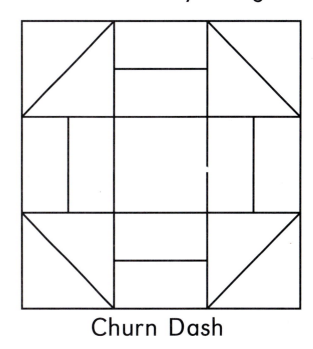

Churn Dash

Which quilt pattern has more?

3. ☐ _____

4. △ _____

 5. Trace and color shapes to make a quilt pattern.

284   two hundred eighty-four

You can put this in your Math Portfolio.

Name _____

Open and Closed Figures

When is a figure open?
When is a figure closed?

Color each closed figure .
✓ each open figure.

1. 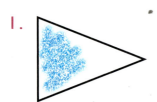 2.  3. 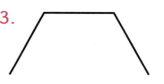 4.

5. (rectangle) 6. 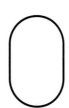 7.  8.

Draw lines to close each.

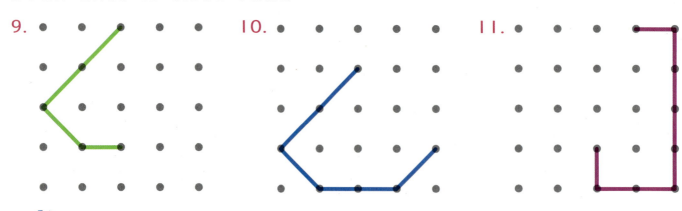

Can you make a closed figure with 2 lines?

Have your child draw a pattern of open and closed figures and explain it.

two hundred eighty-five **285**

Sort the figures. Cut and paste.

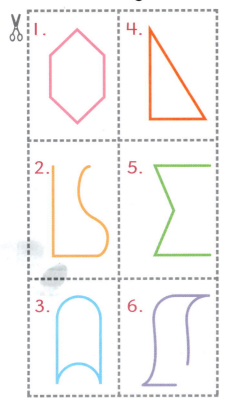

Open Figures	Closed Figures

 7. Janet and Kyle made closed figures. Daryl made an open figure. Kyle used the most ⬬. Which figure did each child make?

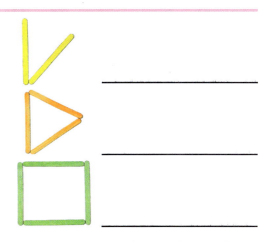

_____

_____

_____

 Draw each.

8. an open figure

9. a closed figure

286   two hundred eighty-six

Name _____

Sides and Corners

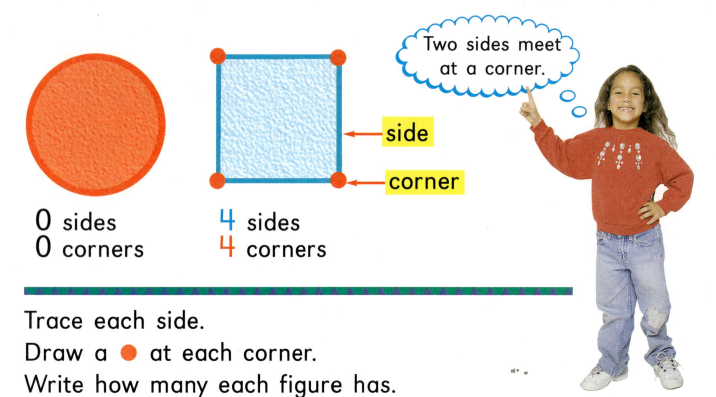

0 sides
0 corners

4 sides
4 corners

"Two sides meet at a corner."

Trace each side.
Draw a ● at each corner.
Write how many each figure has.

1.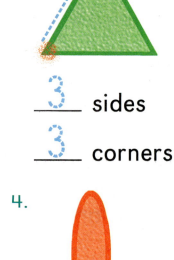

   _3_ sides
   _3_ corners

2.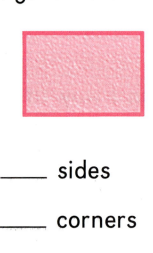

   ___ sides
   ___ corners

3.

   ___ sides
   ___ corners

4.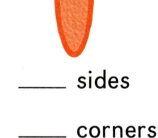

   ___ sides
   ___ corners

5.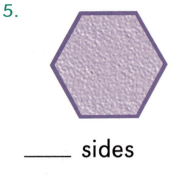

   ___ sides
   ___ corners

6.

   ___ sides
   ___ corners

 What is the pattern in 1–6?

8-2 Draw three different shapes. Have your child tell which shape has the most sides and which the fewest corners.

two hundred eighty-seven **287**

Draw each figure.

1. ☐ with 1 dot inside

2. ⏢ with 0 dots inside

3. ▭ with 2 dots inside

4. △ with 1 dot inside

Is there more than 1 way to draw any figure in 1–4? Tell how to draw it.

Draw a picture. Write Yes or No.

5. Cory drew a figure. The 5 sides met at 5 corners. Is it a closed figure? _____

6. Jenna drew a figure. She drew 4 sides and 3 corners. Is it a closed figure? _____

288 two hundred eighty-eight

Name _____

Plane Figures

Plane figures are flat.

square   triangle   circle   rectangle

 Which figure has no sides and no corners? How are squares and rectangles alike? How are they different?

Ring three figures that have the same shape. Write the name of the shape.

1. 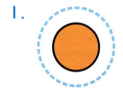    circle

2.   _____

3.   _____

4. 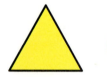   _____

5. 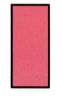    _____

Ask your child to explain why a circle is different from a triangle, square, and rectangle.

Complete each pattern.

1. 2.

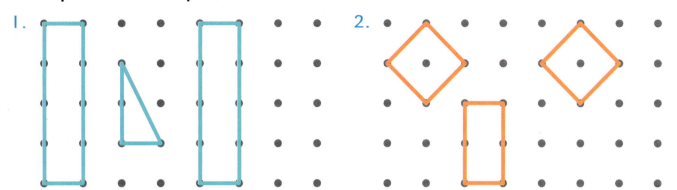

 Andy made this car from plane figures. Make a tally.

3. He used the least number of

   _____.

4. He used one more circle than

   _____.

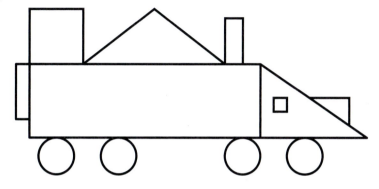

Plane Figure	Tally
square	
triangle	
rectangle	
circle	

5. He used the same number of triangles as

   _____.

 6. Find something you can trace to make each of these plane figures.

square   circle   triangle   rectangle

290   two hundred ninety

Name _____

Write 1 way to buy each book.

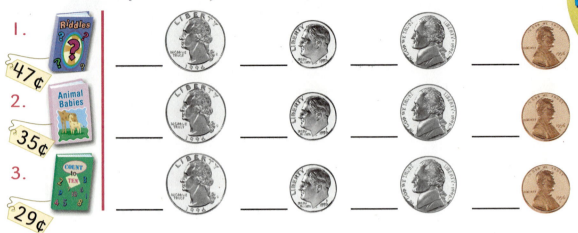

1. 47¢
2. 35¢
3. 29¢

4. Write how much money each child has.
   Color to show which book each can buy.

Match each time with a clock.

5. half past 3 ____
6. 30 minutes after 2 ____
7. four o'clock ____
8. four thirty ____

Write and draw the time.

9. Came at 1:30.
   Stayed 1 hour.
   Left at ____ : ____ .

10. Came at 3:30.
    Stayed a half hour.
    Left at ____ : ____ .

two hundred ninety-one

Same Shape

Name _____

You can find plane figures everywhere.

 The sign has **4 sides** and **4 corners**. It is a **rectangle**.

 Why is the sign not a square?

Trace each figure.

1.
2.
3.
4.
5.

6.
7.
8.
9.
10.

11.
12.
13.
14.

Color 1 ☐ in the graph for each figure in 1–14.

15.

Shape	Number Shown				
square					
circle					
rectangle					
triangle	☐				
	0	1	2	3	4

Name _____

Same Shape Same Size

Are the figures the same shape and the same size?

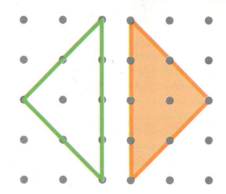

1st Name the figures.

2nd Match the sides exactly.

The triangles are the same shape and size.

 Name another way to check that two figures are the same shape and size.

Color figures with the same shape yellow.
Color red if these figures are the same size.

1.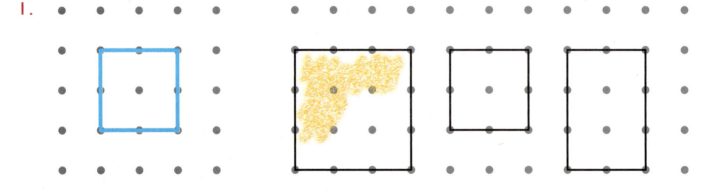

2.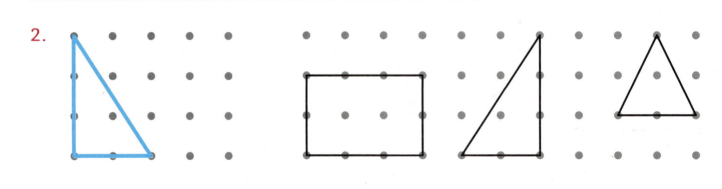

Draw a figure with the same shape and size. Describe and name each.

1. 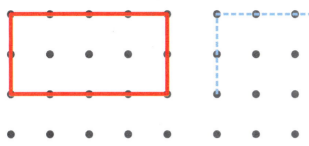 Each has

   _____ sides

   _____ corners

   _____

2. 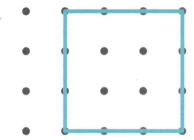 Each has

   _____ sides

   _____ corners

   _____

3. 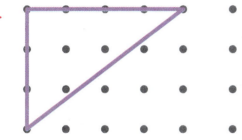 Each has

   _____ sides

   _____ corners

   _____

 How much does each cost?

 3¢    6¢    5¢   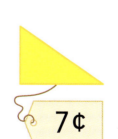 7¢    4¢

4. 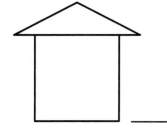 _____

5. 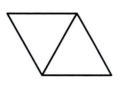 _____

6. 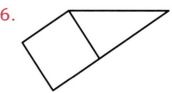 _____

294   two hundred ninety-four

Name _____

Symmetry

Make **matching parts**.

| 1st | Fold and trace. | 2nd | Cut and open. |

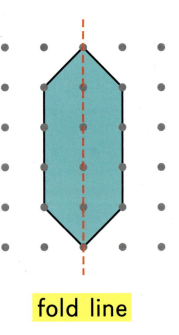

fold line

 Describe the matching parts.
How else can you fold to make matching parts?

Ring the figure that shows matching parts.

1.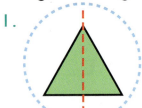
2.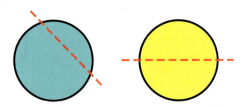

3.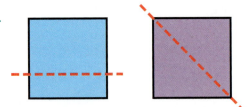
4.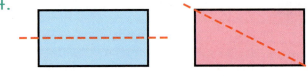

5.
6.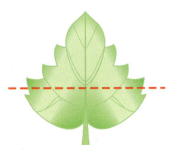

Have your child trace an object he/she can fold to make matching parts.

two hundred ninety-five 295

Draw a line to make matching parts.

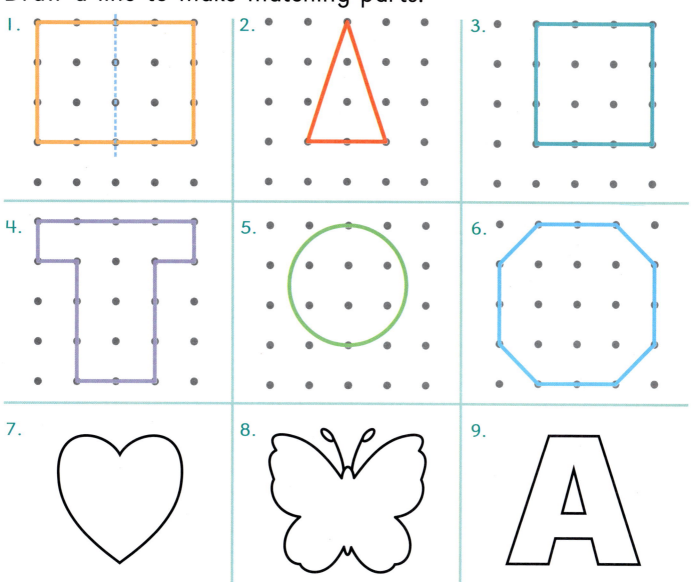

 10. Trace and cut out. Fold in different ways to make matching parts.

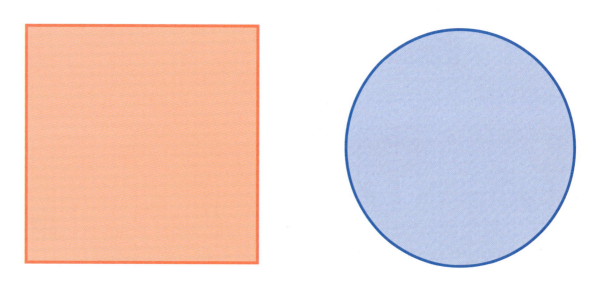

Name _____

Slides and Turns

You can model a slide and a turn.

slide pattern   turn pattern

Color each slide 🖍 blue and each turn 🖍 red.

1.

2.

3.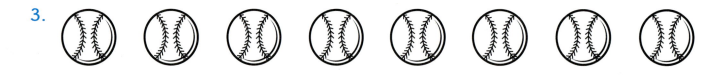

4.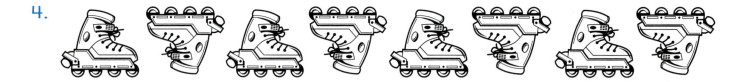

Draw what is next. Ring slide or turn.

5. 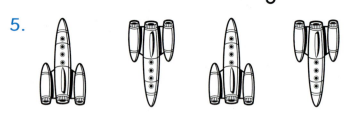   slide or turn

6.    slide or turn

7. Draw a slide pattern and a turn pattern.

two hundred ninety-seven **297**

Solid Figures

Name _____

Solid figures can have curved or flat surfaces.

cone — flat

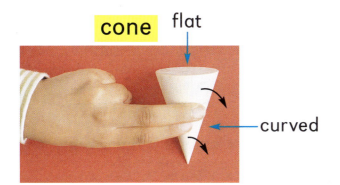

pyramid

— curved

 Which solid can roll? Why?
Which solid can slide? Why?
Why can you call solids space figures?

Color solids that roll .

1.  sphere	2. 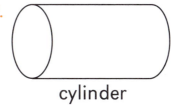 cylinder	3. 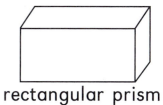 rectangular prism
4. 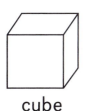 cube	5. 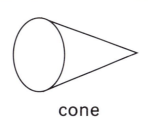 cone	6.  pyramid
7.  rectangular prism	8. 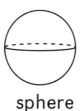 sphere	9. 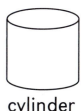 cylinder

 In 1–9 ✓ solids that roll and slide.

 Which solids can you stack:
10. one on top of the other?   11. only on the top?

298 two hundred ninety-eight

Name _____

Cube, Pyramid, Rectangular Prism

These space figures have all
<mark>flat surfaces.</mark>

   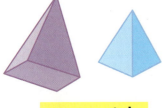   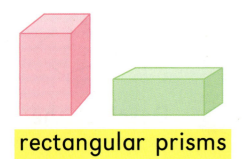

<mark>cube</mark>   <mark>pyramids</mark>   <mark>rectangular prisms</mark>

TALK IT OVER
How are these solids alike?
How are they different?
How many flat surfaces does each have?

---

1. Ring each cube blue, each pyramid red, and each rectangular prism green.

PROBLEM SOLVING
2. How many ways can Tyrell stack 2 cubes, 1 pyramid, and 1 rectangular prism?   _____ ways

 8-9 Ask your child to show you some space figures with flat surfaces that are squares, rectangles, or triangles.

two hundred ninety-nine **299**

Dasil traced the flat surface of a pyramid.

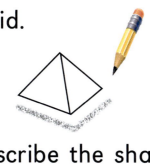

 Describe the shape made. Is its shape the same as the other flat surfaces?

Complete the table.

Shape made	Sides	Corners	Ring the surface traced.
1. ▭ (pink rectangle)	4		🟥 🔺 ▬
2. ◼ (green square)			◼ ▲ ▬
3. ▲ (blue triangle)			◼ ▲ ▬

Color the solid to match each shape made.

4. 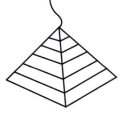   5.    6.    7.

8.    9.    10. 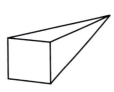   11.

12. Which shape makes both △ and □ ? _____

Name _____

Cylinder, Cone, Sphere

These space figures have **curved surfaces**.

cylinder   cone   sphere

 How are these solids alike? How are they different?

1. Ring each sphere , each cylinder (red), and each cone (green).

 Write cylinder, sphere, or cone.

2. Gary stacks 3 solids that have curved surfaces. Which ones can he stack and how?

top _____

middle _____

bottom _____

8-10 Ask your child to point to solids pictured above that roll, slide, and/or stack.

three hundred one **301**

Which solid can you trace for each figure?

1. Color blue for a circle.
2. Color red for a triangle.

Talk It Over Describe the surfaces you can not trace.

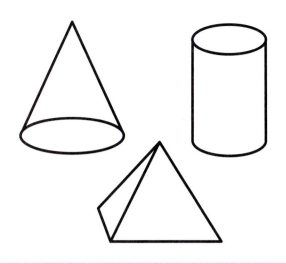

Tally the space figures.
Then write the number for each.

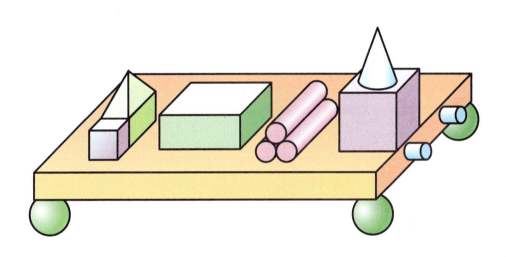

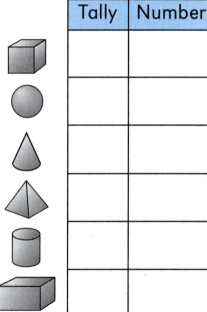

Tally	Number

Match the solid with its name.

3.    cone
   cube
   sphere
   rectangular prism

4.   cylinder
   sphere
   cube
   rectangular prism

5.   pyramid
   cylinder
   cone
   sphere

6.    pyramid
   cylinder
   cone
   cube

7.    sphere
   cylinder
   cone
   rectangular prism

8.   cylinder
   sphere
   cone
   cube

302   three hundred two

Name _____

One Fourth

The spinner shows four equal parts.

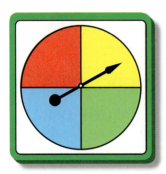

Four equal parts are **fourths**.

Each part is **one fourth**.

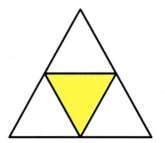

$\dfrac{1 \text{ part yellow}}{4 \text{ equal parts}} = \dfrac{1}{4}$

Ring each shape that shows $\dfrac{1}{4}$ colored.

1.
2.
3.

4.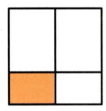
5.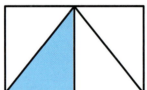
6.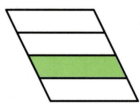
7.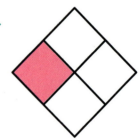

8.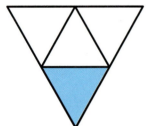
9.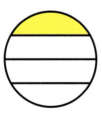
10.
11.

12.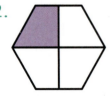
13.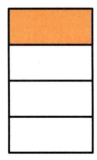
14.
15.

Have your child tell how many fourths are in one whole.

three hundred eleven

Color one fourth. Write about each.

1.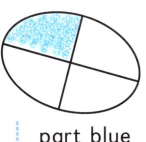
   $\frac{1}{4}$ part blue
   equal parts

2.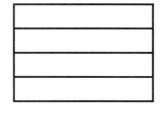
   __ part blue
   __ equal parts

3.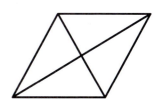
   __ part blue
   __ equal parts

4.
   __ part blue
   __ equal parts

5.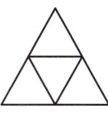
   __ part blue
   __ equal parts

6.
   __ part blue
   __ equal parts

Make 4 equal parts. Color $\frac{1}{4}$.

7.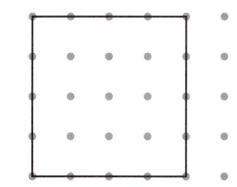

8.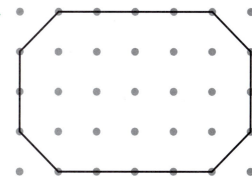

 Which part is left to color?

9. Marta colors 1 part green and 1 part blue out of 3 equal parts.

   _____ left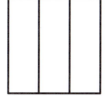

10. Gordon colors 2 parts red and 1 part yellow out of 4 equal parts.

    _____ left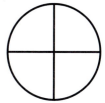

Name _____

Part of a Set

A **fraction** names **part of a set**.

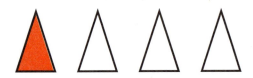

1st	Count how many in all. 4 triangles

$\dfrac{1 \text{ red triangle}}{4 \text{ triangles in all}} = \dfrac{1}{4}$

2nd	Count how many are colored. 1 red triangle

One fourth of the triangles is red.

 Suppose 1 of 2 circles is blue. What part of the set is blue?

What part is colored? Ring.

1.

   $\dfrac{1}{2}$ $\dfrac{1}{3}$ $\dfrac{1}{4}$

2.

   $\dfrac{1}{2}$ $\dfrac{1}{3}$ $\dfrac{1}{4}$

3.

   $\dfrac{1}{2}$ $\dfrac{1}{3}$ $\dfrac{1}{4}$

4.

   $\dfrac{1}{2}$ $\dfrac{1}{3}$ $\dfrac{1}{4}$

5.

   $\dfrac{1}{2}$ $\dfrac{1}{3}$ $\dfrac{1}{4}$

6.

   $\dfrac{1}{2}$ $\dfrac{1}{3}$ $\dfrac{1}{4}$

Give your child 2, then 3, then 4 coins and ask him/her to show one half, one third, and one fourth of a set as heads.

three hundred thirteen **313**

Color part of each set blue .

1. one third 
2. one half 
3. one fourth

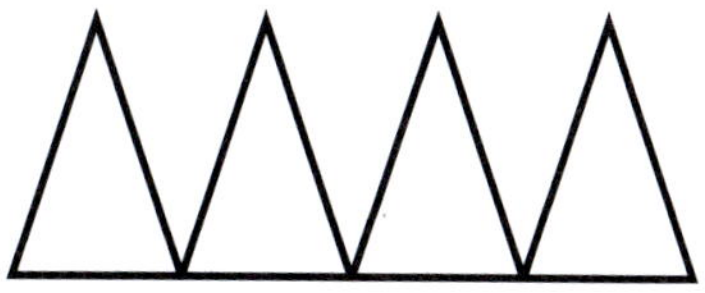

4. one half 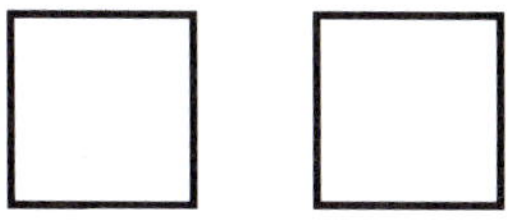
5. one fourth 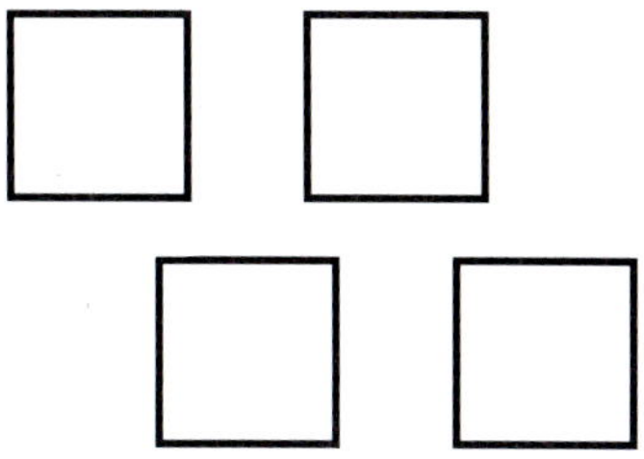
6. one third

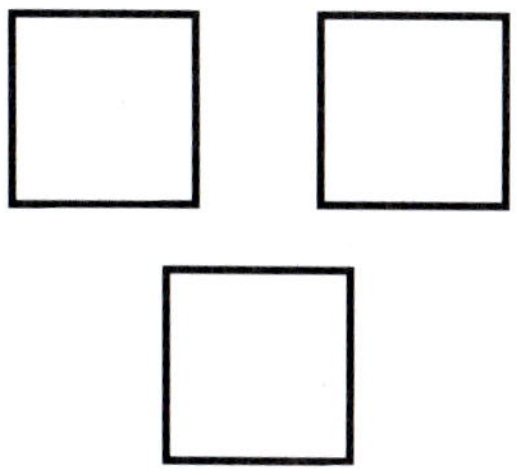

7. one fourth 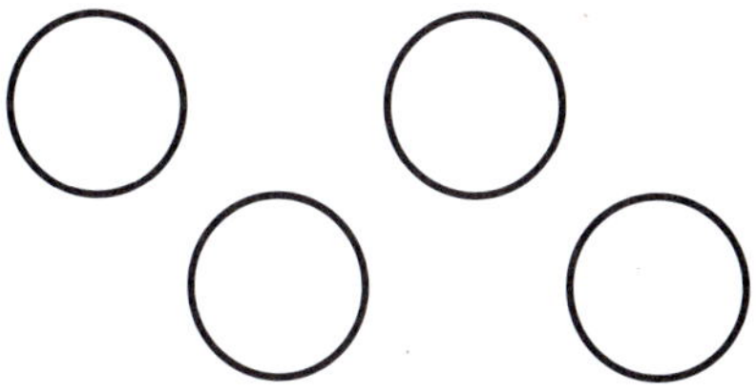
8. one third 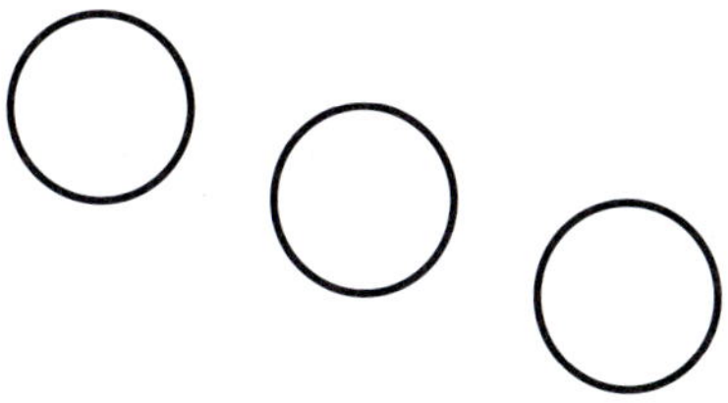
9. one half

PROBLEM SOLVING Draw a picture.
Write your answer in 2 ways.

10. One of 4 ✏ is red. What part of the set is red? _____

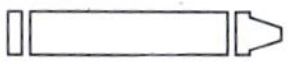

11. One of 2 ⚽ is yellow. What part of the set is yellow? _____

12. One of 3 kites is blue. What part of the set is blue? _____

314 three hundred fourteen

Name _____

More or Less Likely

Gina predicts her spinner is **less likely** to land on red than on yellow or blue. Which color is it **more likely** to land on?

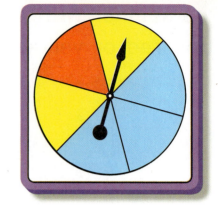

1st	Predict the color.
2nd	Spin. Then tally.
3rd	Write the number.

 Why was the spinner less likely to land on red than on blue?

Color	Tally	Number
red		
blue		
yellow		

Color each spinner to land on

1.  yellow more than red or blue.

2. red less than yellow or blue.

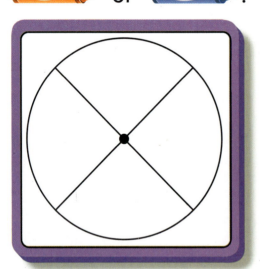

Color	Tally	Number
red		
blue		
yellow		

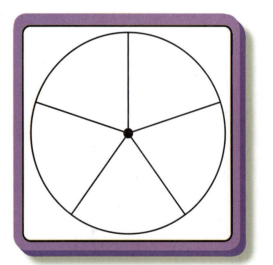

Color	Tally	Number
red		
blue		
yellow		

8-19 Help your child draw a spinner with 8 sections, then ask him/her to color it to land on more blue than red and yellow.

Sometimes there is an **equal chance**.
✔ the spinner with an equal chance of getting red or yellow. Test your answer.

1.

Color	Tally	Number
red		
yellow		

2.

Color	Tally	Number
red		
yellow		

Color each ◯ to show an equal chance of picking blue without looking.

3. 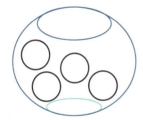 4.  5.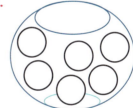

TALK IT OVER How did you know how many marbles to color blue?

FINDING TOGETHER Predict the chance of getting heads or tails. Flip a penny 10 times.

Tally. Write the number.

6.

	Tally	Number
heads		
tails		

320 three hundred twenty

Name _____

This is a LOGO turtle.

Use these commands to move the turtle:

FORWARD FD
RIGHT RT
LEFT LT

To make a square corner, the turtle can turn

RIGHT 90   or LEFT 90

To move the turtle FORWARD 10 steps, press
| F | D | space bar | 1 | 0 | Enter |

To turn RIGHT 90 and move FORWARD 10 steps, press
| R | T | space bar | 9 | 0 | Enter |
| F | D | space bar | 1 | 0 | Enter |

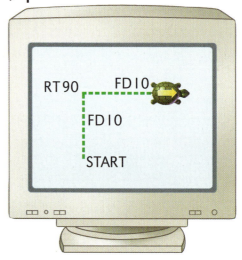

Which way did the turtle turn? Ring.

1.   (RT) LT
   1st 2nd

2.   RT LT
   1st 2nd

3.   RT LT
   1st 2nd

4.   RT LT
   1st 2nd

Which way did the turtle move? Ring.

5. FD 10   LT 90   FD 10

6. FD 10   RT 90   FD 10

In this lesson your child is introduced to the computer language LOGO.

three hundred twenty-one 321

1. Help the turtle get to the lake.

Guess how many  ▪ cover each.
Use your ▪ to check.

2. Guess _____ 
   Check _____

3.  Guess _____ 
   Check _____

322 three hundred twenty-two

Visit Sadlier on the Internet at
www.sadlier-oxford.com

Name _____

PROBLEM-SOLVING STRATEGY
Logical Reasoning

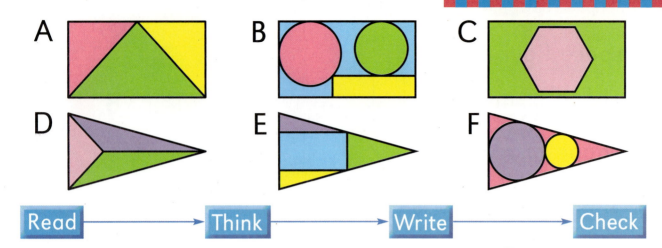

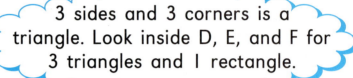

Which flag has:

1. 3 sides and 3 corners; and 3 triangles and 1 rectangle inside?
   A   D   (E)   F

3 sides and 3 corners is a triangle. Look inside D, E, and F for 3 triangles and 1 rectangle.

2. 4 sides and 4 corners; and 2 circles and 1 rectangle inside?
   A   B   E   F

3. 4 sides and 4 corners; and a shape with 6 sides and 6 corners inside?
   A   C   D   E

4. 3 sides and 3 corners; and 2 shapes with no corners inside?
   A   B   D   F

5. 4 sides and 4 corners; and 3 triangles inside?
   A   C   D   E

6. 3 sides and 3 corners; and 3 triangles inside?
   A   C   D   E

Draw a flag with shapes inside. Tell about your flag.

Read → Think → Write → Check

Which space figure am I?

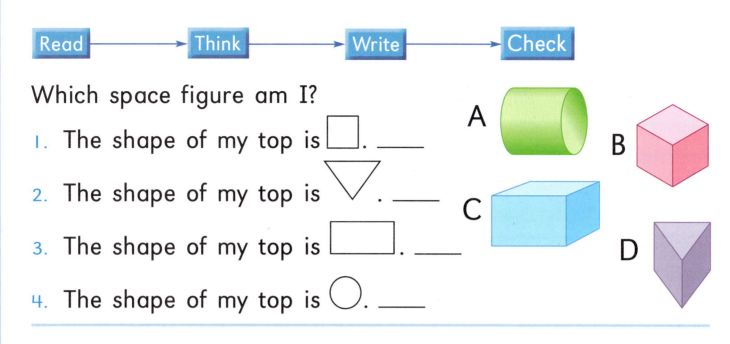

1. The shape of my top is ☐. ____
2. The shape of my top is ▽. ____
3. The shape of my top is ▭. ____
4. The shape of my top is ○. ____

5. How can you stack these figures? Write 1st for bottom, 2nd for middle, and 3rd for top.

 ____   ____  ____

Draw the missing part of each plane figure hidden by ▯.

6. 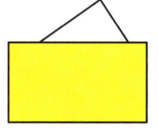   7. 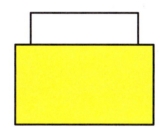   8.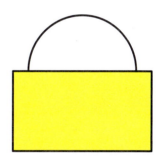

9. Dad cuts an orange in half. He sees 2 flat shapes. Draw what Dad sees.

10. Ann makes this shape. How many 🟧 does Ann use?

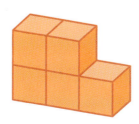 ____

324 three hundred twenty-four

Name _____

PROBLEM-SOLVING APPLICATIONS
Use Drawings and Models

Read ⟶ Think ⟶ Ring ⟶ Check

1. Tina fills 2 shelves of the bookcase.
What part of the bookcase is not filled?
$\frac{1}{3}$ is not filled.

   I of 3 equal parts is not filled.

   $\frac{1}{2}$
   $\frac{1}{3}$
   $\frac{1}{4}$

2. Gina breaks a . What part of it does Gina give away?
Gina gives ____ away.

   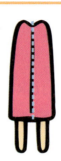

   $\frac{1}{2}$ $\frac{1}{3}$ $\frac{1}{4}$

3. Pedro eats 1 part of a ▢. He gives 2 parts away. What part is left?

   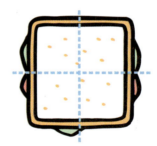

   $\frac{1}{2}$ $\frac{1}{3}$ $\frac{1}{4}$

   ____ is left.

4. Amy ate $\frac{1}{3}$. Daryl ate one third. Is there any pizza left? How much?

   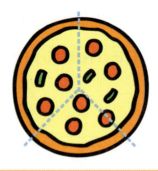

   Yes or No

   ____ is left.

5. Jeff colors one half of the hats [red] and $\frac{1}{2}$ of them [blue]. How many hats are red?

   1 is red.
   2 are red.
   3 are red.

8-22 Ask your child to tell how she/he drew a picture to check problem 3 above.

three hundred twenty-five **325**

Use a strategy you have learned.

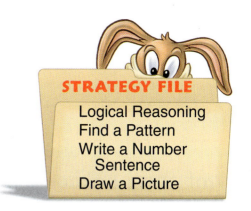

STRATEGY FILE
Logical Reasoning
Find a Pattern
Write a Number Sentence
Draw a Picture

1. Diana folds a ▢ into 4 equal parts. She colors 1 part blue, 1 part green, and 1 part red. Write the fraction for the part left to color.

   ____ left

2. Carly pulls down a shade. It covers $\frac{1}{3}$ of a window. Which window is it? Ring.

3. How can Rob color to make a spinner with 4 equal parts so he is more likely to land on yellow than on red?

   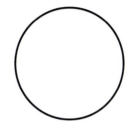

4. Mr. Fox began a pattern of 12 figures. When he finishes, how many figures will show halves?

   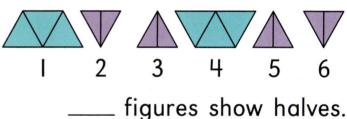

   ____ figures show halves.

5. Find the sum of the numbers inside the ○. ____

6. Find the sum of the numbers not inside the ▢. ____

   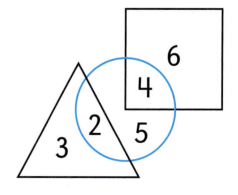

7. Write another problem like 5 or 6.

326 three hundred twenty-six

Chapter Review and Practice

Match the name with the figure.

1. triangle
2. circle
3. rectangle
4. square

Same size and shape? Ring Yes or No.

5.   Yes   No

6.   Yes   No

Match the name, figure, and shape traced.

7. pyramid
8. cone
9. cube
10. rectangular prism

Do the two parts match? Ring Yes or No.

11.   Yes   No

12. 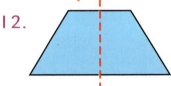   Yes   No

How many equal parts?

13.   ___

14.   ___

What part is shaded? Match.

15.   16.   17.

$\frac{1}{2}$ $\frac{1}{3}$ $\frac{1}{4}$

18. One third of Lee's hats are red. The rest are blue. Color his hats.

19. Color so you are more likely to spin green than red.

three hundred twenty-seven   **327**

Name _____

Connect corners to make other figures inside.

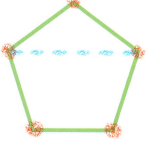

Now there are 2 figures inside.

One figure has __3__ corners.

The other has __4__ corners.

Connect corners to make plane figures inside.

1. 3 triangles
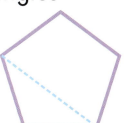

2. 4 triangles

3. 2 triangles

4. 3 triangles

5. 1 square and 1 triangle
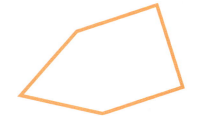

6. 2 triangles and 1 rectangle

7. How many different figures can you make inside this plane figure?

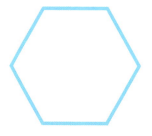

328 three hundred twenty-eight

This page extends your child's understanding of the parts of plane figures.

Performance Assessment

1. Nina traced each flat surface of these solids. Name which solid she used if she made:

   - 6 rectangles.
   - 2 circles.
   - 6 squares.
   - 1 square and 4 triangles.

2. Use dot paper. Draw a figure that has the same shape and size. Draw a different fold line.

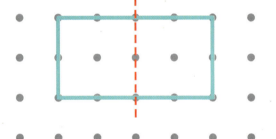
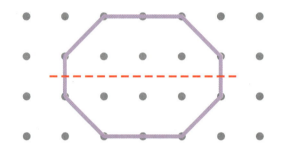

Portfolio

Choose 1 project. Use a separate sheet of paper.

3. Make a fraction book for the different ways to show one third and one fourth.

4. How were these figures sorted? Draw another figure for each circle.

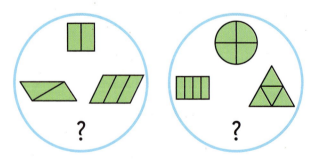

Draw a diagram to sort solids that roll and solids with all flat surfaces.

Check Your Mastery

Name _____

Color to match.

3 sides
4 corners

Solids that roll
Only flat surfaces

1.

2.

Match same shapes. Write the letter.

A   B   C   D

E   F   G   H

3. ___   ___   ___

4. ___ ___ ___

How many equal parts?

5. ___ 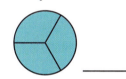 ___

Color the correct rectangle.

6. $\frac{1}{2}$  $\frac{1}{3}$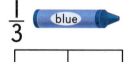

Write Always, Sometimes, Never.

7.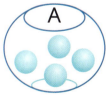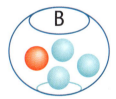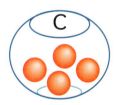

_____ get ●.   _____ get ●.   _____ get ●.

8. What part of the marbles in B is red? ____

 9. Josh folded his paper in half. Ring his paper.

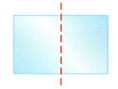 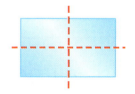

10. Rita is more likely to spin purple. Ring her spinner.

Add and Subtract Two-Digit Numbers

9

CRITICAL THINKING

My school band has 1 more trombone than tuba. It has between 2 and 6 tubas. Which school band is mine?

Math Alive at Home

Dear Family,

Today your child began Chapter 9. As he/she studies addition and subtraction with 2-digit numbers, you may want to read the poem below, which was read in class, with him or her. Encourage your child to talk about some of the math ideas shown on page 331.

Look for the 🏠 at the bottom of each skills lesson. The suggestion on the page gives you an opportunity to improve your child's understanding of math.

You may want to have pennies, dimes, and ▭▭▭▭▭▭▭▭▭▭ available for your child to use throughout this chapter.

For more information about Chapter 9, visit the Family Information Center at **www.sadlier-oxford.com**

Home Activity

Swinging Ten Bands

Try this activity with your child. Make 9 bundles of ten straws. Each bundle of ten is a group of band members. Have your child use the bundles to illustrate addition of tens, as the example below shows. After your child finishes each lesson in this chapter, adjust the operation (addition or subtraction) and groups of band members.

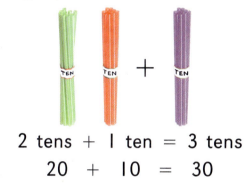

2 tens + 1 ten = 3 tens
20 + 10 = 30

Home Reading Connection

Ten Tom-Toms.

Ten tom-toms,
Timpani, too,
Ten tall tubas
And an old kazoo.

Ten trombones—
give them a hand!
The sitting-standing-marching-running
Big Brass Band

Unknown

MATH CONNECTIONS
Number-Line Estimation

Name _____

Follow directions to show how each band marches.

1.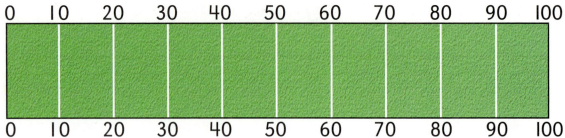

   The band marches:
   forward 40, back 30, forward 30.
   Where is the band on the field? _____

2.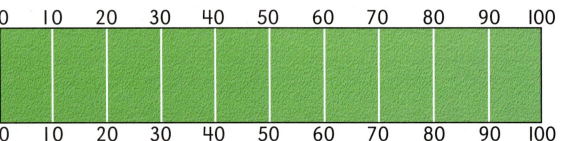

   The band marches:
   forward 80, back 50, back 10.
   Where is the band on the field? _____

3.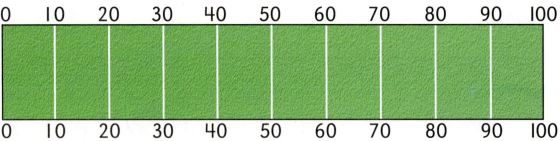

   The band marches:
   forward 80, forward 20, back 10.
   Where is the band on the field? _____

 Which band is closest to 50?
Which band is closest to 100?

 You can put this in your Math Portfolio.

three hundred thirty-three **333**

CROSS-CURRICULAR CONNECTIONS
Multicultural Music

Name _____

Draw strings on each instrument.
Write how many strings you draw in all.

mandolin

ukulele

banjo

lute

5. Survey 6 friends about which instrument they like the best. Show your results on a bar graph.

Name _____

Estimate about How Many

Are there about 20 marchers in this parade?

An *estimate* tells about how many.

First estimate. Then count to check.

1. About how many 🔔 are there? Ring.

          20   30   40

About how many 🥁 are there? Ring.

2.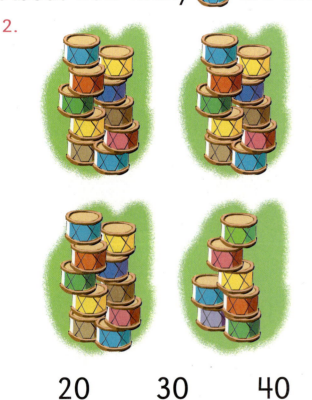

3.

  20   30   40   20   30   40

 9-1 Ask your child to put a counter on any number on the game board pictured on page 336 and to tell whether the counter is closer to 40 or 50.

three hundred thirty-five   **335**

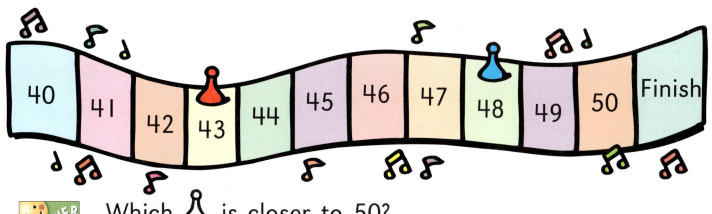

 Which ♟ is closer to 50?

Which ♟ is closer to 40?

Is each number closer to 40 or 50?

1. Put a ▪ on 41.

   It is closer to _____.

2. Put a ▪ on 49.

   It is closer to _____.

3. Put a ▪ on 46.

   It is closer to _____.

4. Put a ▪ on 44.

   It is closer to _____.

 Use a number line.

5. Ben has 78 points.
   Alma has 87 points.
   Who has closer to
   90 points?

   _____

   Who has closer to
   50 points?

   _____

6. Felix has 48 points.
   Elena has 38 points.
   Who has closer
   to 40 points?

   _____

 Make a ▭▭▭ from 60 to 70.

7. Draw a ♟ closer to 60.

8. Draw a ♟ closer to 70.

Add Tens

1 one + 4 ones = 5 ones

$$\begin{array}{r}1\\+4\\\hline 5\end{array}$$

1 ten + 4 tens = 5 tens

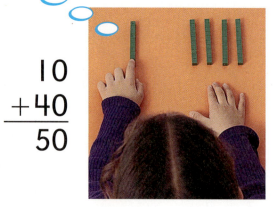

$$\begin{array}{r}10\\+40\\\hline 50\end{array}$$

TALK IT OVER Name other addition sentences that have a sum of 5 tens. Use ▬▬▬ to check.

Add. Use ▬▬▬ to check.

1. 50 + 30 = __80__
   5 tens + 3 tens = __8__ tens

2. 30 + 60 = ___
   3 tens + 6 tens = ___ tens

3. 40 + 20 = ___

4. 20 + 60 = ___

5. 80 + 10 = ___

6. 70 + 20 = ___

7. 10 + 50 = ___

8. 10 + 30 = ___

9. 30 + 40 = ___

10. 50 + 40 = ___

PROBLEM SOLVING

11. Carla has 50 🎟. Malia has 80 🎟. How many more 🎟 does Carla need to have as many as Malia?

    50 + ___ = 80

    ___ more

12. The band has 10 🎺, 20 🎷, and 30 🎷. How many instruments is that in all?

    ___ + ___ + ___

    ___ in all

9-2 Provide ▬▬▬ for your child to add 10 to 20, 40, and 60 and to describe the pattern.

Subtract Tens

Name _____

7 ones − 4 ones = 3 ones   7 tens − 4 tens = 3 tens

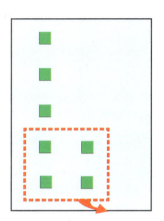
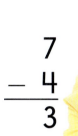

"To subtract tens, I think of ones."

$$\begin{array}{r} 7 \\ -\ 4 \\ \hline 3 \end{array} \qquad \begin{array}{r} 70 \\ -\ 40 \\ \hline 30 \end{array}$$

Find the difference. Use ▭▭▭ to check.

1. 4 tens − 2 tens = __2__ tens
   40 − 20 = __20__

2. 5 tens − 2 tens = ____ tens
   50 − 20 = ____

3. 60 − 30 = ____   4. 80 − 40 = ____

5. 20 − 10 = ____   6. 90 − 80 = ____

7. 70 − 60 = ____   8. 30 − 10 = ____

9. 90 − 50 = ____   10. 70 − 50 = ____

11. 60 − 40 = ____   12. 50 − 10 = ____

TALK IT OVER 30 + 40 = 70 is the related addition sentence for 70 − 40 = 30. Name the related addition sentences for 1–12.

338   three hundred thirty-eight

Ask your child to show how 80 − 50 is like 8 − 5.

Name _____

More Adding Tens and Ones

I have 23 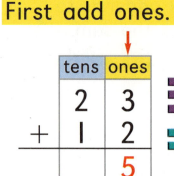 and 12 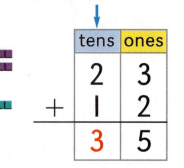. How many is that altogether?

23 + 12 = ___?

35 altogether

First add ones.

tens	ones
2	3
+1	2
	5

Then add tens.

tens	ones
2	3
+1	2
3	5

Add. Use ▭▭▭▭ and ▫ to check.

1.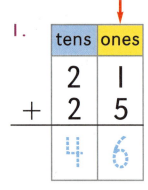

tens	ones
2	1
+2	5
4	6

2.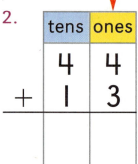

tens	ones
4	4
+1	3

3. 11   41   34   37   63   24
   +66   +54   +24   +22   +25   +52

4. 89   56   81   53   72   12
   +10   +21   +17   +26   +21   +53

5. 42   74   40   32   65   70
   +55   +13   +34   +67   +33   +19

6. 50 + 16 = ___   43 + 20 = ___   31 + 40 = ___

 7. In 6 what happens when one addend has 0 ones?

9-6 Invite your child to show why 23+34 is the same as 34+23.

three hundred forty-three **343**

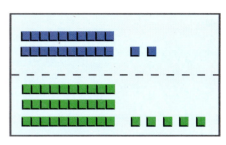

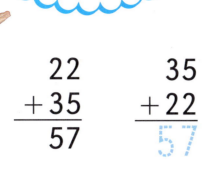

Change the order of the addends to check the sum.

```
  22        35
+ 35      + 22
----      ----
  57        57
```

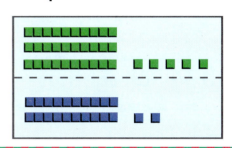

equals

Find the sum. Change the order to check.

1.
```
  58      31       32        16
+ 31    + 58     + 42      + 43
```

2.
```
  63             36         42
+ 22           + 62       + 56
```

3. Kyle has 34 music books. His brother has double that number. How many books does his brother have?   _____ books

Add. Write the next additions in the pattern.

4.
```
  21      32      43      54
+ 12    + 12    + 12    + 12     + ___     + ___
```

344   three hundred forty-four

Name _____

Add Tens or Ones

Ruiz has 32 stickers.
He gets 20 more.
Anita has 32 stickers.
She gets 2 more.
Who has fewer stickers?

Anita has fewer.

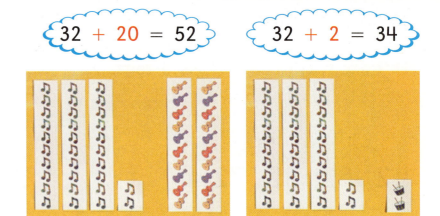

$32 + 20 = 52$ $32 + 2 = 34$

 TALK IT OVER How is adding 2 tens and 2 ones different? How would you write each in an addition frame?

 MENTAL MATH Add mentally. Write the sum.

1. 25
   + 3

    28

2. 31
   +20
   ----

3. 74   80   43   62   51   24
   + 2   + 4   + 3   + 6   + 7   + 5

4. 35   46   29   20   30   50
   +40   +50   +30   +65   +52   +36

5. 29   42   68   81   72   37
   +60   + 7   + 1   +10   +20   + 2

9-7 Say a number between 20 and 25. Have your child count on to add 2 tens and then count on to add 2 ones.

three hundred forty-five 345

You can **count on** by tens or by ones to add mentally.

42 + 30   *30 is 3 tens.*   85 + 3   *3 is 3 ones.*

   +10 +10 +10   +1   +1   +1
42, 52, 62, 72   85, 86, 87, 88

So 42 + 30 = 72.   So 85 + 3 = 88.

Add mentally. Watch for tens or ones.

1. 71 + 20 = ___
   71 + 2 = ___

2. 34 + 40 = ___
   34 + 4 = ___

3. 15 + 30 = ___
   15 + 3 = ___

4. 16 + 30 = ___
   16 + 3 = ___

5. 24 + 3 = ___
   24 + 30 = ___

6. 75 + 2 = ___
   75 + 20 = ___

7. 28 + 40 = ___

8. 92 + 2 = ___

9. 35 + 4 = ___

 Why are the sums in 1 not equal?

---

 10. Fill in each ☐.

17 + ☐ = 19   23 + ☐ = 33   62 + ☐ = 65

☐ + 30 = 75   ☐ + 2 = 38   ☐ + 3 = 56

Name _____

Subtract Tens and Ones

45 − 23 = __?__

First subtract ones.

tens	ones
4	5
− 2	3
	2

Then subtract tens.

tens	ones
4	5
− 2	3
2	2

The difference is 22.

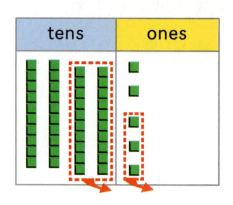

Ring the ▬▬▬ and ■ you subtract. Write the difference.

1.
tens	ones
5	8
− 3	5
2	3

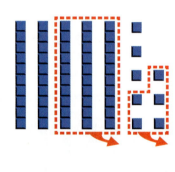

2.
tens	ones
2	7
− 1	2

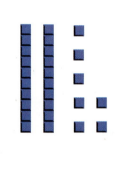

3.
tens	ones
2	6
− 1	4

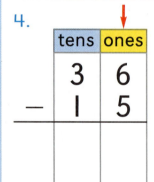

4.
tens	ones
3	6
− 1	5

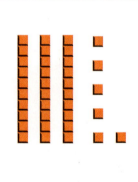

5.
tens	ones
3	4
− 1	1

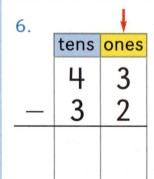

6.
tens	ones
4	3
− 3	2

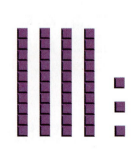

9-8 Ask your child to pick any exercise on this page and explain how to solve it.

three hundred forty-seven 347

Write the whole and the part subtracted.
First subtract ones. Then subtract tens.

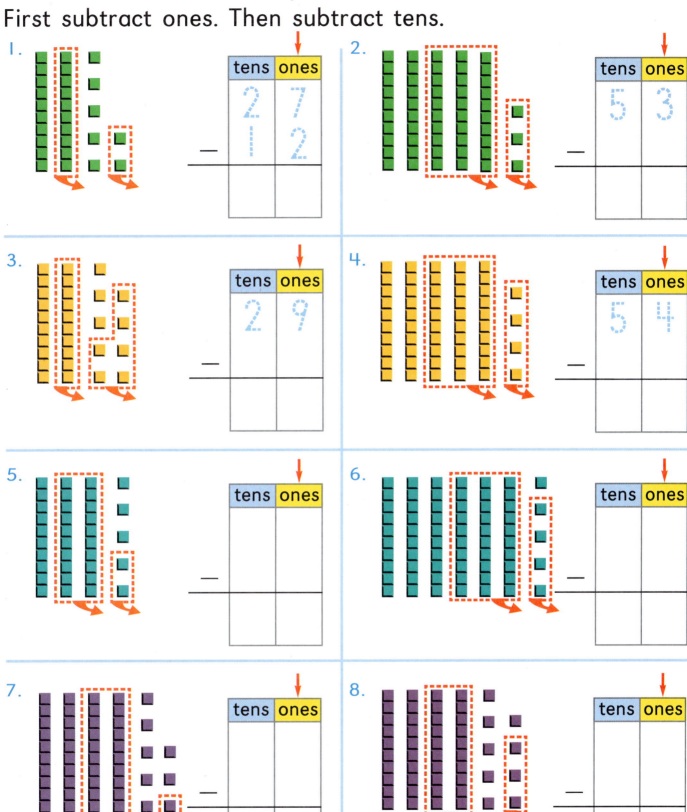

9. Subtract 1 ten 1 one from: 41, 51, 61, 71, 81.
10. Subtract 2 tens 2 ones from: 63, 53, 43, 23.

Name _____

More Subtracting Tens and Ones

A band has 45 flags.
14 are red.
How many are not red?

45 − 14 = __?__

31 flags are not red.

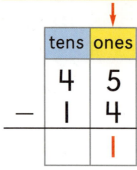
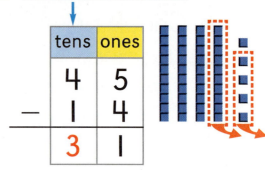

Subtract. Use ▭▭▭▭▭ and ▪ to check.

1. 58
   −16

2. 32
   −11

3. 75   67   84   37   98   79
   −12   −23   −13   −32   −51   −47

4. 51   96   56   88   83   59
   −41   −54   −13   −44   −61   −19

5. 49   69   47   99   68   76
   −34   −38   −25   −36   −45   −32

 TALK IT OVER What happens when you subtract all of the tens or all of the ones?

33 − 3 tens = __?__
33 − 3 ones = __?__

Ask your child to make up a subtraction problem using the numbers 56 and 24.

9-9

three hundred forty-nine **349**

Check by adding.

```
 46  in all              22  part
-24  part subtracted    +24  part
 22  part left           46  in all
```

Find the difference. Add to check.

1.
   55   23   78   87
  -32   +32   -63   -44
   23

2.
   39   57   69
  -15   -21   -45

PROBLEM SOLVING Use a model.

3. Val and Pete each had 65 .
   Val lost 23. Pete lost 25.
   Who has more left?   _____ has more left.

4. Dee has between 36 and
   39 . She lost 12 .
   Then she had 26 .
   How many did Dee
   have at first?   _____

36, ?, ?, 39

350 three hundred fifty

Name _____

Subtract Tens or Ones

Liz makes 35 cards.
She sends 20.
Raul makes 35 cards.
He sends 2. Who has
more cards left?

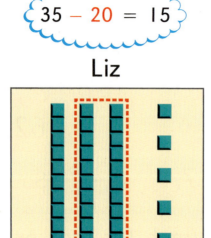

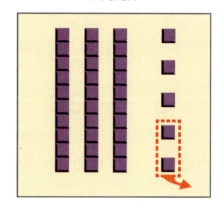

Raul has more.

TALK IT OVER How is subtracting 2 tens and 2 ones different? Tell how to write each in a subtraction frame.

 Subtract mentally. Write the difference.

1. 36
   − 5
   ─────
   31

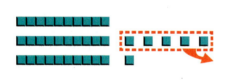

2. 47
   −20
   ─────

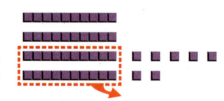

3. 59   78   48   95   84   39
   − 8   − 5   − 2   − 4   − 2   − 7
   ───   ───   ───   ───   ───   ───

4. 19   45   67   82   58   79
   −10   −20   −30   −40   −40   −50
   ───   ───   ───   ───   ───   ───

5. 74   29   33   38   59   58
   − 4   −10   −20   − 4   − 6   −30
   ───   ───   ───   ───   ───   ───

9-10 Have your child explain what happens in subtraction when the numbers in the ones place are the same.

three hundred fifty-one 351

You can count back by tens or by ones to subtract mentally.

57 − 30   *30 is 3 tens.*

−10 −10 −10
27, 37, 47, 57

So 57 − 30 = 27.

65 − 3   *3 is 3 ones.*

−1 −1 −1
62, 63, 64, 65

So 65 − 3 = 62.

Subtract mentally. Watch for tens or ones.

1. 44 − 20 = 24
   44 − 2 = 42

2. 66 − 30 = ___
   66 − 3 = ___

3. 28 − 10 = ___
   28 − 1 = ___

4. 59 − 40 = ___
   59 − 4 = ___

5. 35 − 2 = ___
   35 − 20 = ___

6. 83 − 3 = ___
   83 − 30 = ___

7. 29 − 2 = ___

8. 72 − 30 = ___

9. 56 − 4 = ___

  Why are the differences in 1–3 not equal?

  Spin for second number. Write each number sentence.

10. 56 − ___ = ___

11. 77 − ___ = ___

12. 32 + ___ = ___

13. 53 + ___ = ___

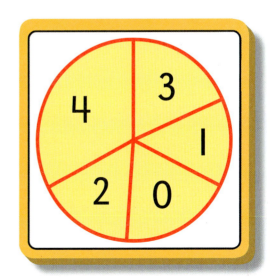

Name _____

1. Write always, sometimes, or never for what you would pick.

Bag A
open figure

Bag B
plane figure

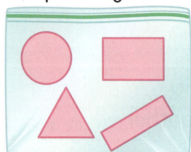

Bag C
solid with flat surface

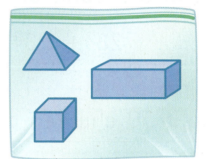

_____   _____   _____

2. What part of the group of figures in Bag B has 3 corners? _____

3. What part of the group of solids in Bag C can you trace to make a triangle? _____

4. Write more or less for what you are likely to pick.

Bag D
equal parts

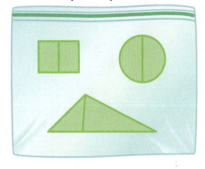

Bag E
thirds

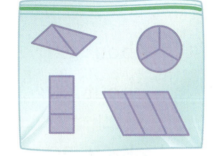

Bag F
fourths

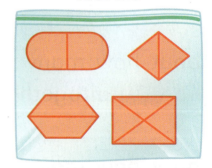

_____   _____   _____

5. Draw the figures from Bag D that have symmetry.

This page reviews the mathematical content presented in Chapter 8.

three hundred fifty-three **353**

Estimating Sums and Differences

Name _____

There are 28 boys and 23 girls in the band. About how many children are in the band altogether?

28 is close to 30
23 is close to 20

$$28 + 23 = \underline{?}$$

Estimate $30 + 20 = 50$
about 50 children altogether

 How can you estimate the difference between the number of boys and girls?

Estimate the answer. Use a .

1. $41 + 52$ is about $\underline{90}$.
2. $79 - 27$ is about ____.
3. $29 + 64$ is about ____.
4. $62 - 18$ is about ____.

Use the table. Ring about how many.

Band Instruments

Instrument	Count
(wood block)	14
(bongos)	18
(kazoo)	37
(tambourine)	23

5. About how many more 🎵(kazoo) than 🥁(bongos)?
   about 20
   about 30

6. About how many 🎵(wood block) and 🥁(tambourine) altogether?
   about 30
   about 40

7. About how many fewer 🎵(wood block) than 🎵(kazoo)?
   about 20
   about 30

8. About how many in all?   about 80   about 90

354 three hundred fifty-four

Have your child use the data in the table to make up addition and subtraction problems involving estimation.

9-11

Name _____

Add Money

22¢ + 12¢ = ___?___

dimes	pennies
🪙 🪙	🪙 🪙
🪙	🪙 🪙

22¢ + 12¢ = 34¢

Add pennies.

dimes	pennies
2	2
+ 1	2
	4

Then add dimes.

dimes	pennies
2	2
+ 1	2
3	4

Use 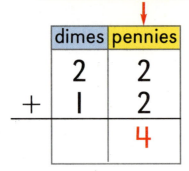 and 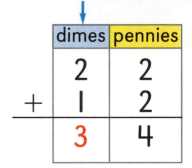. Add.

1.
dimes	pennies
2	8
+ 3	1
5	9

28¢
+31¢
59¢

2.
dimes	pennies
3	3
+ 4	1

33¢
+41¢

Find the sum.

3. 16¢   40¢   62¢   32¢   10¢
   +51¢   +19¢   +37¢   +26¢   +40¢

4. 26¢   34¢   35¢   59¢   16¢
   +50¢   +35¢   +52¢   +40¢   +72¢

5. 41¢   52¢   23¢   37¢   43¢
   +26¢   +17¢   +32¢   +51¢   +23¢

6. How is adding dimes and pennies like adding tens and ones?

Place 9 dimes and 9 pennies in a bag. Have your child grab 2 handfuls, write the addition sentence, and tell the total amount.

three hundred fifty-five **355**

Use 🪙 and 🪙 to buy both items.
How much money is spent in all?

1.

   17¢ + 31¢ = ____

   17¢
   +31¢
   48¢

2.

   ____ + ____ = ____

3.

   ____ + ____ = ____

4.

   ____ + ____ = ____

TALK IT OVER Tell how many dimes and pennies you spent each time in 1–4.

CRITICAL THINKING Find 2 different ways to spend 75¢.

A 32¢   B 25¢   C 43¢   D 50¢

5. ____ and ____

   ____ + ____ = 75¢

6. ____ and ____

   ____ + ____ = 75¢

356 three hundred fifty-six

Subtract Money

You can subtract money using dimes and pennies.

Subtract pennies.
Then subtract dimes.

dimes	pennies
3	2
− 1	2
2	0

$$\begin{array}{r} 32¢ \\ -12¢ \\ \hline 20¢ \end{array}$$

32¢ − 12¢ = 20¢

Use 🪙 and 🪙. Subtract.

1.
dimes	pennies
4	6
− 3	4
1	2

$$\begin{array}{r} 46¢ \\ -34¢ \\ \hline 12¢ \end{array}$$

2.
dimes	pennies
5	7
− 1	3

$$\begin{array}{r} 57¢ \\ -13¢ \\ \hline \end{array}$$

Find the difference.

3. 44¢   89¢   78¢   96¢   86¢
   −21¢ −45¢ −25¢ −72¢ −55¢

4. 58¢   99¢   65¢   73¢   99¢
   −37¢ −10¢ −35¢ −50¢ −72¢

 Write subtraction stories to show each:

5. losing all the 🪙.

6. losing all the .

Use 🪙 and 🪙. Find the change.

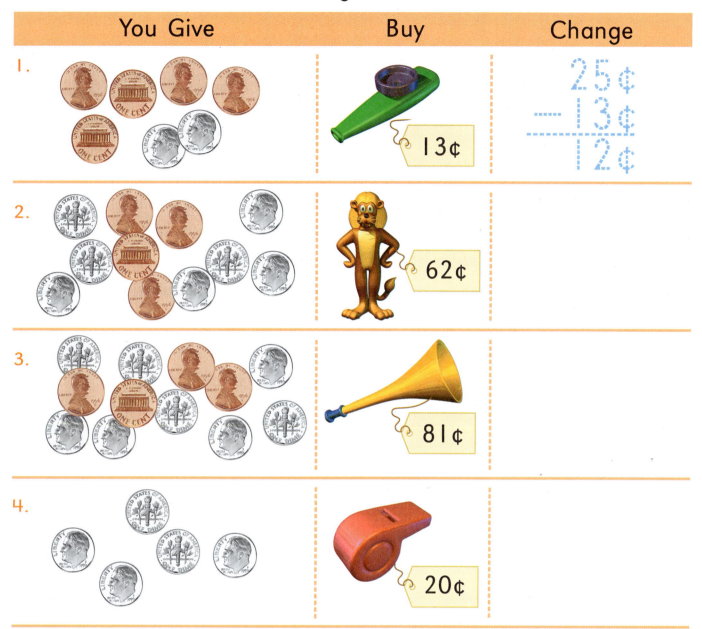

	You Give	Buy	Change
1.		13¢	25¢ −13¢ 12¢
2.		62¢	
3.		81¢	
4.		20¢	

PROBLEM SOLVING Use and .

5. A 🥁 costs 44¢. Lena has 95¢ Can she buy 2 🥁?

6. Yan has 45¢. A 🚩 costs 65¢. How much more money does Yan need to buy it?

CRITICAL THINKING

7. I bought a 🎵 for 58¢. Draw 2 other ways to pay 58¢.

Name _____

Using Addition and Subtraction

What does 18 + 10 − 8 equal?
Work from left to right.

First add.
18 + 10 = 28

Then subtract.
28 − 8 = 20

18 + 10 − 8 = 20

MENTAL MATH

Ring the part you do first.
Add or subtract mentally.

1. (65 − 20) + 3 = 48
   45 + 3

2. 82 − 40 + 5 = ___

3. 49 − 5 + 50 = ___

4. 23 − 2 + 30 = ___

5. 36 + 10 − 4 = ___

6. 54 + 30 − 1 = ___

7. 57 − 6 − 20 = ___

8. 95 − 40 − 3 = ___

MAKE UP YOUR OWN

9. ___ ◯ ___ ◯ ___ = ___

Have your child explain how he/she solved the exercises above mentally.

three hundred fifty-nine **359**

Add or subtract. Watch for + and −.

1.
$$\begin{array}{r}34\\+24\\\hline 58\end{array}\quad\begin{array}{r}46\\+20\\\hline\end{array}\quad\begin{array}{r}62\\+35\\\hline\end{array}\quad\begin{array}{r}54¢\\+15¢\\\hline\end{array}\quad\begin{array}{r}36¢\\+42¢\\\hline\end{array}\quad\begin{array}{r}13¢\\+44¢\\\hline\end{array}$$

2.
$$\begin{array}{r}67\\-43\\\hline 24\end{array}\quad\begin{array}{r}39\\-17\\\hline\end{array}\quad\begin{array}{r}77\\-26\\\hline\end{array}\quad\begin{array}{r}48¢\\-26¢\\\hline\end{array}\quad\begin{array}{r}56¢\\-25¢\\\hline\end{array}\quad\begin{array}{r}99¢\\-43¢\\\hline\end{array}$$

3.
$$\begin{array}{r}32\\+27\\\hline\end{array}\quad\begin{array}{r}69\\-23\\\hline\end{array}\quad\begin{array}{r}43\\+35\\\hline\end{array}\quad\begin{array}{r}13¢\\+31¢\\\hline\end{array}\quad\begin{array}{r}95¢\\-43¢\\\hline\end{array}\quad\begin{array}{r}87¢\\-54¢\\\hline\end{array}$$

 SECOND LOOK

4. Ring sums between 50 and 60 in 1–3.

5. ✔ even differences.

 PROBLEM SOLVING Use a model to solve each problem.

6. Nina and Flo each have 25¢.
   Nina spends 12¢.
   Flo spends 10¢.
   Who has more money left?

   _____ has more money left.

7. Mario has 27¢. He finds 12¢. Then Mario loses 15¢. How much does Mario have now?

   Mario has ____ now.

8. Sue and Rob have the same number of stickers. Sue gives 4 away. Who has more stickers now?

   _____ has more.

9. Chico has 8 sets of cards. He loses 3 sets. Then he finds 2 of the lost sets. How many sets does he have now?

   ____ sets

Name _____

Regroup in Addition

2 tens 6 ones + 5 ones = __?__

Add the ones.

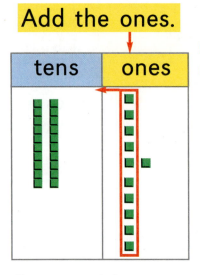

You can regroup 10 ones as 1 ten.
11 ones = 1 ten 1 one

Add the tens.

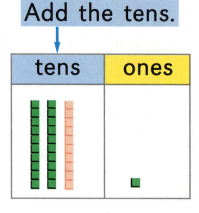

2 tens 11 ones

3 tens 1 one

Add the ones. Regroup when you can.

1. 5 tens 9 ones + 4 ones

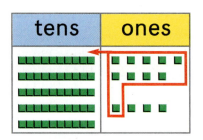

5 tens 13 ones = ___

6 tens 3 ones = ___

2. 3 tens 5 ones + 5 ones

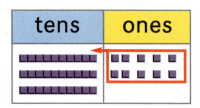

3 tens ___ ones = ___

___ tens ___ ones = ___

3. 4 tens 7 ones + 5 ones

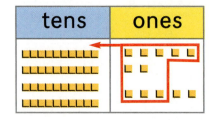

4 tens ___ ones = ___

___ tens ___ ones = ___

4. 1 ten 4 ones + 8 ones

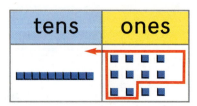

1 ten ___ ones = ___

___ tens ___ ones = ___

9-15 Have your child show different ways to make 1 ten from a total of 10 ones— for example, 1 one and 9 ones.

three hundred sixty-one **361**

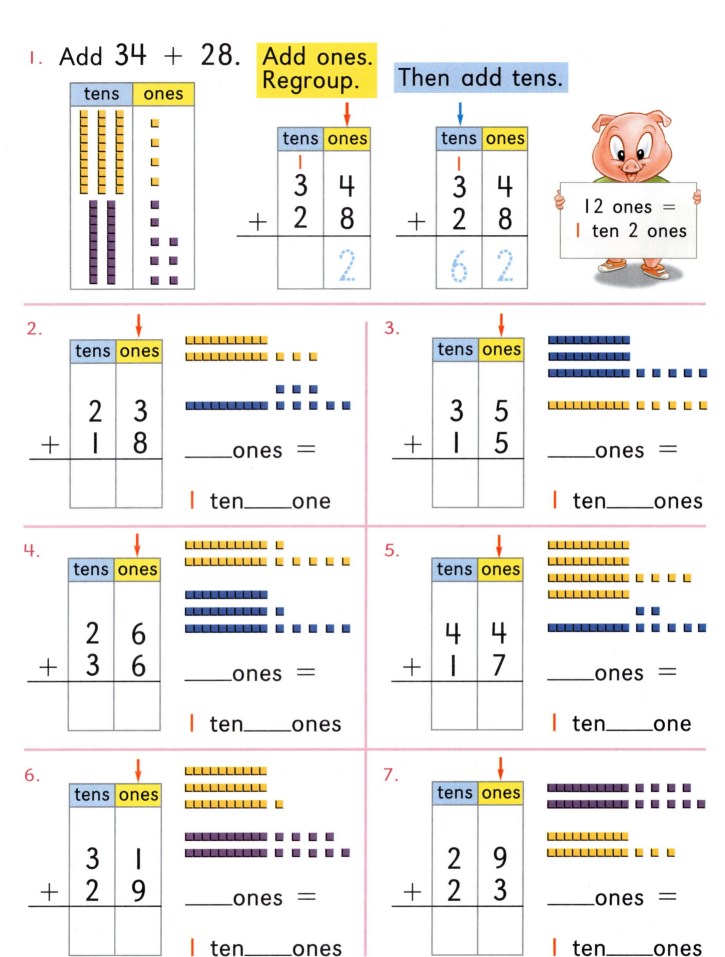

When can you regroup ones?

Name _____

Regroup in Subtraction

3 tens 1 one − 3 ones = ?

Need to regroup ones.

Subtract ones. Then subtract tens.

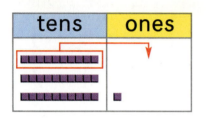

Regroup 1 ten as 10 ones. Now you can subtract 3 ones.

3 tens 1 one = 2 tens 11 ones

2 tens 8 ones

Subtract after you regroup.

1. 2 tens 2 ones − 9 ones =

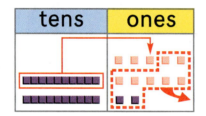

1 ten _12_ ones − 9 ones

= ___ ten ___ ones

2. 2 tens 1 one − 6 ones =

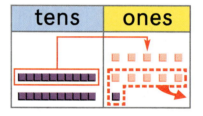

_1_ ten ___ ones − 6 ones

= ___ ten ___ ones

3. 4 tens 1 one − 4 ones =

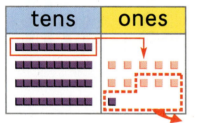

_3_ tens ___ ones − 4 ones

= ___ tens ___ ones

4. 3 tens 0 ones − 7 ones =

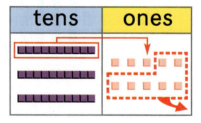

_2_ tens ___ ones − 7 ones

= ___ tens ___ ones

5. Dee has 20 stickers. She uses 8 stickers and gives away 3. How many stickers are left?   ___ left

Subtract. Regroup. Ring to take away. **12 ones − 7 ones**

1. 32 − 17 = ?

tens	ones
3	2
− 1	7

Not enough ones, so regroup.

3 tens 2 ones
= **2** tens **12** ones

2 tens − 1 ten

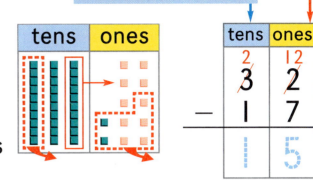

tens	ones
²⁄3	¹²⁄2
− 1	7
	5

2.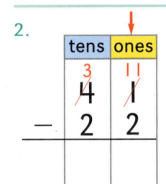

tens	ones
³⁄4	¹¹⁄1
− 2	2

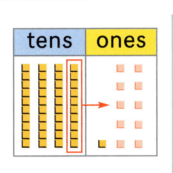

4 tens 1 one
= *3* tens *11* ones

3.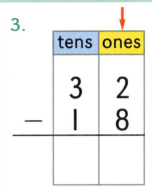

tens	ones
3	2
− 1	8

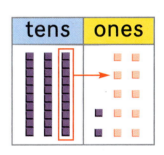

3 tens 2 ones
= ___ tens ___ ones

4.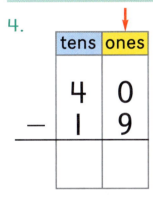

tens	ones
4	0
− 1	9

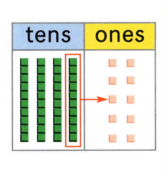

4 tens 0 ones
= ___ tens ___ ones

5.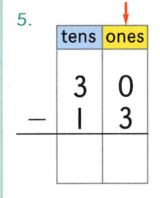

tens	ones
3	0
− 1	3

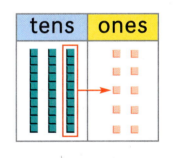

3 tens 0 ones
= ___ tens ___ ones

6.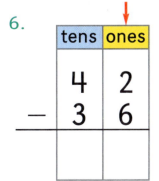

tens	ones
4	2
− 3	6

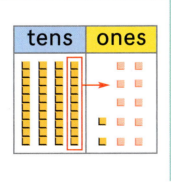

7.

tens	ones
2	1
− 1	5

Name _____

PROBLEM-SOLVING STRATEGY:
Logical Reasoning

Use the clues to solve each problem.

First write how much money each bank has.

Bank A 37¢

1. It has more than 4 coins.
   It has less than 40¢.

   ~~A~~ B C ~~D~~

   Bank __C__ has ____ ¢.

2. It has less than 5 coins.
   It has no pennies.

   A B C D

   Bank ____ has ____ ¢.

Bank B ____

3. It has more than 37¢.
   It has only 1 nickel.

   A B C D

   Bank ____ has ____ ¢.

Bank C ____

4. It has between 35¢ and 40¢.
   It has no nickels.

   A B C D

   Bank ____ has ____ ¢.

Bank D ____

5. Make up different clues for Bank C.

Have your child make up different clues for Bank A.

three hundred sixty-five **365**

Solve using logical reasoning.   Read → Think → Write → Check

1. It has between 40¢ and 50¢.
   It has less than 6 coins.

   ~~E~~ F G ~~H~~ ~~I~~

   Bank ____ has ____.

   E ____

2. It has more than 45¢.
   It has more than 2 pennies.

   E F G H I

   Bank ____ has ____.

   F ____

3. It has more than 45¢.
   It has more than 2 dimes.

   E F G H I

   Bank ____ has ____.

   G ____

4. It has less than 47¢.
   It has more than 5 coins.

   E F G H I

   Bank ____ has ____.

   H ____

5. It has more than 50¢.
   It does not have 7 coins.

   E F G H I

   Bank ____ has ____.

   I ____

6. Use your coins to show 39¢.
   Give 2 clues. Have a partner
   name the coins you used.

PROBLEM SOLVING

366 three hundred sixty-six

Name _____

PROBLEM-SOLVING APPLICATIONS:
Choose the Operation

Read → Think → Write → Check

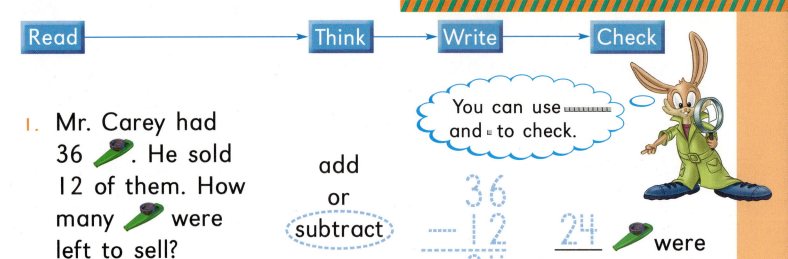

1. Mr. Carey had 36 🎵. He sold 12 of them. How many 🎵 were left to sell?

   add or ~~subtract~~

   36
   −12
   ─────
   24

   __24__ 🎵 were left to sell.

2. Mrs. Cruz bought 63 🎵. She bought 13 more. How many 🎵 in all?

   add or subtract

   ____  in all

3. There were 45 🧍 in the band. 14 more joined. How many 🧍 are in the band now?

   add or subtract

   ____ 🧍 in the band now.

4. The band had 96 🎵. 34 were red. How many 🎵 were not red?

   add or subtract

   ____  were not red.

TALK IT OVER How can you change the order to check addition problems? How can you add to check subtraction problems?

In this lesson your child used counters to check addition and subtraction problems.

three hundred sixty-seven **367**

Read → Think → Write → Check

1. I had 4 dimes 3 pennies. I found 25¢ more. How much money do I now have?

   add or subtract

   I have ____.

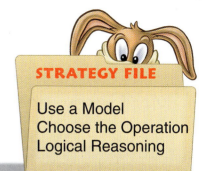

STRATEGY FILE

Use a Model
Choose the Operation
Logical Reasoning

2. I had 38¢. I lost 2 dimes 6 pennies. How much money do I have left?

   add or subtract

   I have ____ left.

3. Sue had 5 dimes 6 pennies. She spent 23¢. Does she have an equal number of dimes and pennies left?

   Yes or No

4. One of my 6 coins is a quarter and one is a penny. I have the same number of dimes as nickels. How much money do I have?

   I have ____.

5. I had 34¢. My mom gave me a quarter. Then I spent 42¢. How much money do I have now?

   I have ____ now.

6. Use these coins. Write a money problem.

Name _____

Chapter Review and Practice

Add or subtract. Use ▭▭▭ and ▫.

1. 53   34   25   76   61   81
   +36   +44   +14   +23   +10   + 6
   ---   ---   ---   ---   ---   ---
   89

2. 54   49   98   69   77   89
   −42   −47   −83   − 9   −31   −64
   ---   ---   ---   ---   ---   ---

3. 89   48   96   67   23   38
   −73   +21   −64   −45   +75   +41
   ---   ---   ---   ---   ---   ---

 ✔ the sums that have more than 7 tens in 1–3.

Use 🪙 and 🪙. Add or subtract.

4. 67¢   27¢   23¢   66¢   84¢
   −20¢   +51¢   +61¢   −35¢   −53¢
   ----   ----   ----   ----   ----

5. 79¢   33¢   25¢   78¢   42¢
   −68¢   +34¢   +61¢   −44¢   +55¢
   ----   ----   ----   ----   ----

6. A band has 59 🚩. 29 of them are put away. About how many are left?

   About ____ 🚩 left

7. A band has 21 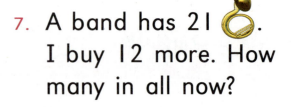. I buy 12 more. How many in all now?

   ____  in all

Name _____

Add or subtract.
Use the code to answer the riddle.
Why was the baker sad?

Code:

A	B	C	D	E	G	H	I	O	S	T	U
26	68	70	10	47	40	55	23	96	63	57	36

25 +43	98 −51	30 +40	89 −63	31 + 5	51 +12	45 + 2	87 −32	22 +25
68								
B								

31 +24	14 +12	80 −70	51 + 6	23 +73	99 −31	69 −22	22 + 4	89 −32

50 + 5	48 −25	61 + 2	13 +34	60 −20	10 +30	88 −25

370 three hundred seventy

This page extends your child's understanding of two-digit addition and subtraction.

Name _____

Inches

This is 1 **inch**. |—1 inch—|

Measure like this. Line up the ruler with the object.

Measure inches with an **inch ruler**.

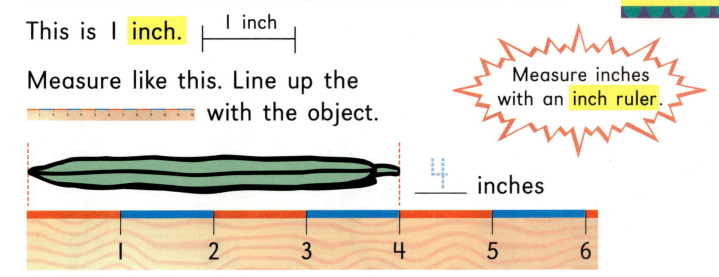

4 inches

 Name something that is about 1 inch.

Use your ruler. Measure the length.

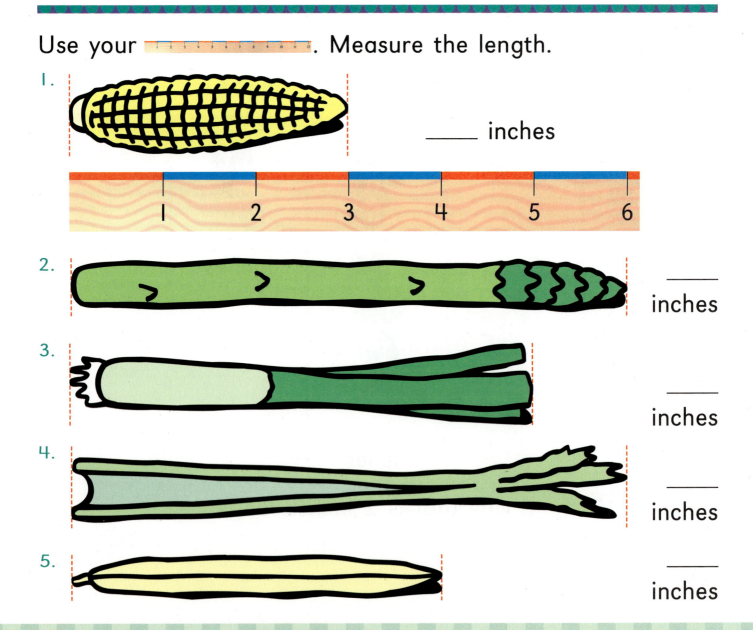

1. ____ inches
2. ____ inches
3. ____ inches
4. ____ inches
5. ____ inches

Use your _____.
Measure the height in inches.

1. _____ inches

2. _____ inches

3. _____ inches

4. _____ inches

5. _____ inches

 6. Draw a vine that is 2 inches longer.

384 three hundred eighty-four

Name _____

Feet

A ▬▬▬ is **1 foot** long.

There are **12 inches** in 1 foot.

My math book is about 1 foot long.

Estimate the length of each real object.
Then measure with a ▬▬▬. Ring.

1.    (more than 1 foot)

   less than 1 foot

2. [shoe image]   more than 1 foot

   less than 1 foot

3.    more than 1 foot

   less than 1 foot

4.    more than 1 foot

   less than 1 foot

10-4 Direct your child as she/he traces each family member's footprint and uses a ruler to tell if it is more, less, or about 1 foot.

three hundred eighty-five **385**

Color each object that is more than 1 foot.

FINDING TOGETHER

1. List 2 real objects that are about 1 foot.

Name _____

Cups and Pints

I pint holds 2 cups

2 cups = 1 pint

1 pint = 2 cups

Which holds more? Ring.

1.  or

2.  or

3.  or

4.  or

5.  or

6.  or

7. Color how many cups.

three hundred eighty-seven **387**

Name ______________________________

2 pints = 1 quart

I quart holds 2 pints.

I quart = 2 pints

Which holds more? Ring

1.  or 

2.  or 

3.  or 

4.  or 

5.  or 

6.  or 

CRITICAL THINKING

Order from least to greatest.
Write 1st, 2nd, and 3rd.

7.

pint ____ quart ____ cup ____

Give your child time to describe how he/she solved the *Critical Thinking* above.

Name _____

Pound

1 pound

less than 1 pound   more than 1 pound

1. ✔ about 1 pound.
   Ring less than 1 pound.

2. ✔ about 1 pound.
   Ring more than 1 pound.

10-7   Tell your child to find something in the kitchen that is about 1 pound — for example, a full box of macaroni or a full loaf of bread.

three hundred eighty-nine   **389**

Ring the heaviest. ✔ the lightest.

1.

2.

 In 1 and 2, is the heaviest object greater than 1 pound? Is the lighest less than 1 pound?

3. Order from lightest to heaviest. Write 1st, 2nd, and 3rd.

 ____    ____    ____

 Draw each object where it belongs.

4.    5.

390   three hundred ninety

Name _____

Centimeters

This is 1 **centimeter**.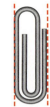

Measure with a **centimeter ruler**.

Line up the ruler with the object.

The **width** of a large paper clip is about 1 centimeter.

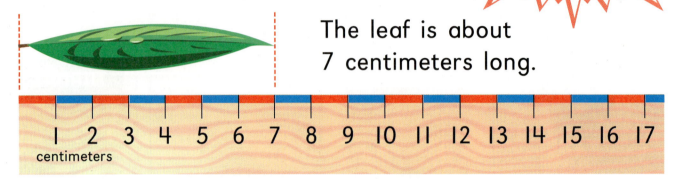

The leaf is about 7 centimeters long.

Use your ruler. Measure the length.

1.

   _____ centimeters

2.

   _____ centimeters

3.

   _____ centimeters

4.

   _____ centimeters

5.

   _____ centimeters

Using the ruler on this page, ask your child to measure her/his favorite toy and tell if it is more than or less than 15 centimeters.

three hundred ninety-one

Use your [ruler]. Measure.
Write how many centimeters high each is.

1. 2. 3. 4.

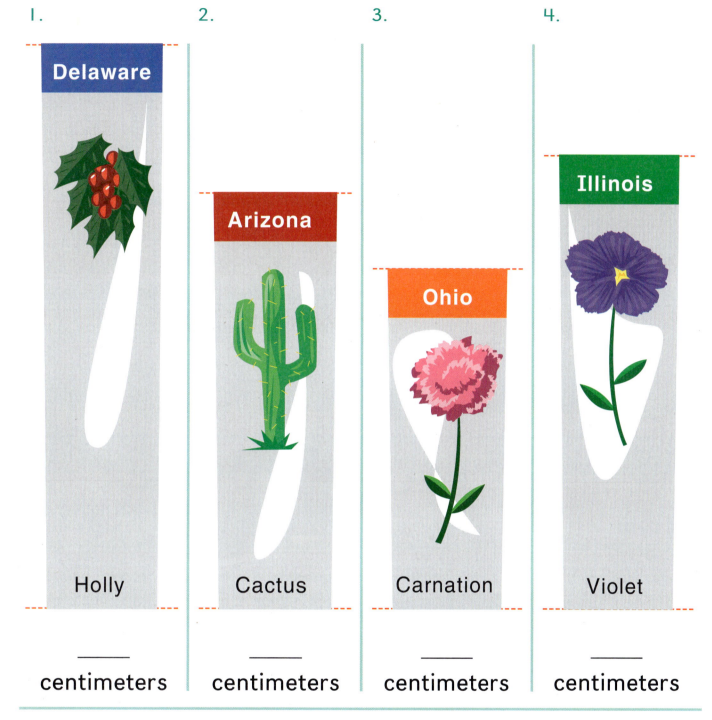

Holly | Cactus | Carnation | Violet

___ centimeters | ___ centimeters | ___ centimeters | ___ centimeters

5. Tyler's bookmark is 16 cm high. Lori's is 13 cm high. How many centimeters higher is Tyler's than Lori's? ___ centimeters

6. Make a bookmark for your state that is between 10 and 20 centimeters high.

Name _____

Using Centimeters

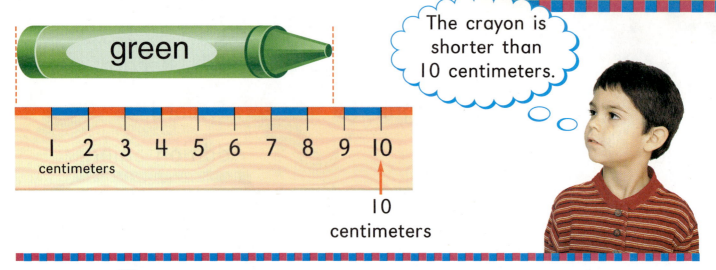

Connect 🟦 to make a train about 10 centimeters.
Ring the object if it is longer than 10 centimeters.

 ✔ the object if it is shorter than 10 centimeters.

 Draw an object that is exactly 10 centimeters long.

Direct your child to measure each family member's foot width and tell which is closest to 10 centimeters.

three hundred ninety-three **393**

Name _____

Spin 2 times to get the top number.

1st First spin for ▬▬▬ or 🪙. Record.

2nd Spin again for ▪ or 🪙. Record.

Add.

Addition Spinner

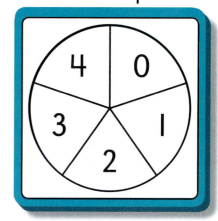

1. ☐☐   ☐☐
   +2 2 + 5

2. ☐☐   ☐☐
   +1 4 +3 3

3. ☐☐¢   ☐☐¢   ☐☐¢   ☐☐¢
   +3 0¢ +4 1¢ +2 4¢ +5 2¢

Subtract.

Subtraction Spinner

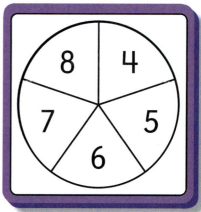

4. ☐☐   ☐☐
   −4 2 − 5

5. ☐☐   ☐☐
   −3 1 −1 3

6. ☐☐¢   ☐☐¢   ☐☐¢   ☐☐¢
   −4 0¢ −2 3¢ − 2¢ −1 1¢

Name _____

Estimate a Liter

1 liter

less than 1 liter

more than 1 liter

 If you were thirsty could you drink 1 liter of water? Would 1 liter fill a bathtub? Would 1 liter fill a paper cup?

Color more than 1 liter .

1.
2.
3.
4.
5.
6.
7.
8.

 In 1–8 color less than 1 liter orange.

 10-10 Prompt your child to list objects from home that could hold more than 1 liter.

three hundred ninety-five **395**

Ring yes or no.

1. holds less than 1 liter?

yes   no

2. holds more than 1 liter?

yes   no

3. holds less than 1 liter?

yes   no

 Ring.

4. Rosa needs 1 liter of water to make punch. Which holds about 1 liter?

5. Mike used more than 1 liter of water to fill the birdbath. Which holds more than 1 liter?

6. Lee made fruit punch in this container. Did he make about 1 liter?

yes

no

7. Naomi drank the juice in this container. Did she drink less than 1 liter?

yes

no

Name _____

Kilogram

less than 1 kilogram   1 kilogram   more than 1 kilogram

Which are more than 1 kilogram? Ring.

Which are less than 1 kilogram? Ring.

Have your child tell whether a full box of his/her favorite snack is more or less than a kilogram. Hint: a kilogram is about 2 pounds.

three hundred ninety-seven   **397**

2 loaves = 1 kilogram

Ring.

1.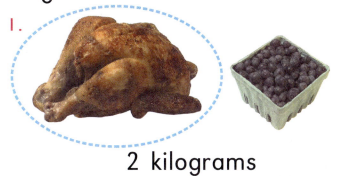

   2 kilograms

2.

   3 kilograms

3.

   1 kilogram

4.

   2 kilograms

5. Order from heaviest to lightest. Write 1st, 2nd, 3rd.

    ____   ____    ____

 1 kilogram = 4 🍎.

6. Color  how many 🍎 are about 2 kilograms.

7. Color  how many 🍎 are less than 1 kilogram.

Name _____

PROBLEM-SOLVING STRATEGY
Use a Map

Solve. Use the map and these steps. •———• is 1 block.

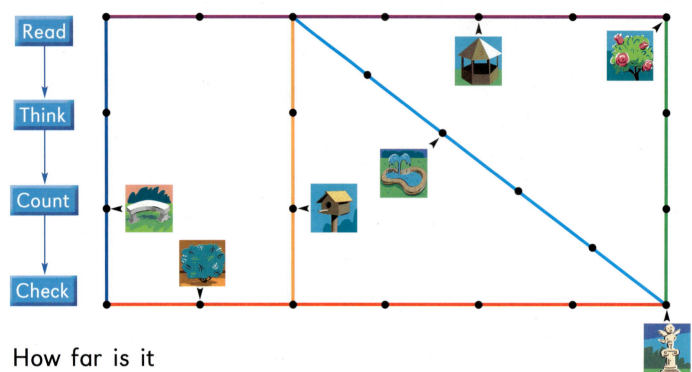

How far is it

1. from to the on the purple path? _2_ blocks

2. from to on the blue path? ____ blocks

3. from to the on the red path? ____ blocks

Use the shortest path. How far is it

Use the **scale**.

4. from the to ? ____ blocks

5. from to the ? ____ blocks

6. from to the ? ____ blocks

7. from to ? ____ blocks

Pick 2 points on the map shown and ask questions such as those above for your child to answer.

four hundred three **403**

Solve. Use the map below.

Use the shortest path.

1. Go from the 🎟️FLOWER SHOW to the 🌼.   __6__ units

2. Go from the 🌻 to the 🌟.   ____ units

3. Go from the 🌿 to the 🌹.   ____ units

Use these paths to go from the 🌼 to the 🌹.

4. green and orange   ____ units

5. green, red, blue, and purple   ____ units

Go from the 🌿 to 🌼.

6. Tell the path you used. Tell what flowers you saw.

7. How far did you go?   ____ units

Start at the 🎟️FLOWER SHOW. Walk to a flower. Tell the story.

404   four hundred four

Name _____

PROBLEM-SOLVING APPLICATIONS
Logical Reasoning

Read → Think → Write → Check

1. Rene's fruit is about 1 pound.
   Carlo's fruit is more than 1 kilogram.
   T.J.'s fruit is the lightest.
   Which fruit does
   each have?

   ____

   ____   ____

2. Mia's jar holds 1 liter.
   Kerry's jar holds 1 pint.
   Julie's jar holds the most.
   Which jar belongs to each girl?

   A   B   C

   Mia ____   Kerry ____   Julie ____

3. Gina's seed is less than
   7 centimeters.
   It is about 1 inch.
   Which is Gina's seed? ____

   It is ____ centimeters long.

   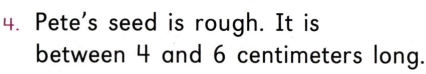

4. Pete's seed is rough. It is
   between 4 and 6 centimeters long.

   Which is Pete's seed? ____

   It is ____ inches long.

   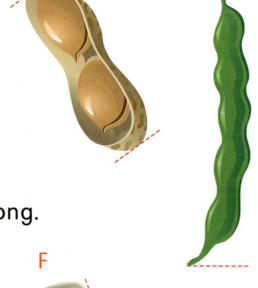

5. Teri's seed is the longest.
   It is more than 7 centimeters.

   Which is Teri's seed? ____   It is ____ inches long.

PROBLEM SOLVING

Read → Think → Write → Check

Use a strategy you have learned.

STRATEGY FILE
Write a Number Sentence
Guess and Test
Logical Reasoning
Draw a Picture

1. At 8 o'clock the temperature was 60°F. At noon it was 10° hotter. At 3:30 it was 10° hotter. What was the temperature at 3:30?

   It was ____°F.

2. Paula made a square flower box. How many 🟩 around is the box?

   It is ____ 🟩 around.

3. Tia needs to make 8 quarts of punch. She has 1 quart of 🍋 juice. Color the bottles she can use.

   8 = _____

4. Match each child with the tool. Joe's 🎃 weighs 1 pound. Tia's 🍆 is 6 inches long. Ben squeezed a glass of juice from an 🍊.

Joe   Tia   Ben

Name _____

Chapter Review and Practice

Use your rulers. Measure.

1.

____ inches

2.

____ centimeters

Which holds less? Ring.

3.  or

4.  or

✔ less than 1 foot

5.

✔ about 1 pound

6.

✔ more than 1 kilogram

7.

✔ more than 1 liter

8.

four hundred seven 407

Name _____

Use 🧊 to build each shape.
Write how many 🧊 used.

1.

_____

2.

_____

3.

_____

4.

_____

5.

_____

6.

_____

7. In 1–2 which shape uses more 🧊?

_____

8. In 1–6 which shape uses the most 🧊?

_____

9. In 1–6 which shape uses the least 🧊?

_____

10. In 1–6 which shapes use the same number of 🧊?

_____

Performance Assessment

Name _____

1. Measure each in inches and in centimeters.

   a chain of 2 large   a train of 5

   ___ in.   ___ cm   ___ in.   ___ cm

2. Color 1 ▢ for each inch.

     ▢▢▢▢▢▢▢▢▢▢

     ▢▢▢▢▢▢▢▢▢▢

   Color 1 ▢ for each centimeter.

     ▢▢▢▢▢▢▢▢▢▢

     ▢▢▢▢▢▢▢▢▢▢

3. Ellen draws a square.

   Each side is 5 ▢ long.   It is about ___ ▢ around.

 Choose one of these projects. Use a separate sheet of paper.

4. Make a map with dot paper. First put 2 places on the map. Then draw 1 long path and 1 short path between the places.

   The long path is ___ •—•.

   The short path is ___ •—•.

5. Make a poster. Draw or cut out pictures to show some uses for each measuring tool.

ruler	cup	scale

This page provides a variety of informal assessment opportunities in order to measure your child's understanding of Chapter 10.

four hundred nine

Check Your Mastery

Name _____

Use your ruler. Measure.

1.

_____ inches

2.

_____ centimeters

Complete.

3. I will fill _____ .

4. I will fill _____ .

5. Write more or less.

1 foot	1 pound	1 liter	1 kilogram

_____	_____	_____	_____

Ring the best tool to use.

6.

7.

8. Ring the temperature. Color the thermometer.

 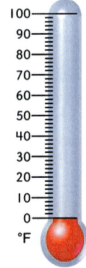

30°F   70°F

Addition and Subtraction Facts to 18

11

CRITICAL THINKING

If Eric is putting his sixth shell into the pail, how many shells are already there?

For more information about Chapter 11, visit the Family Information Center at www.sadlier-oxford.com

Dear Family,

Today your child began Chapter 11. As she/he studies addition and subtraction facts 13 to 18, you may want to read the story below, which was read in class. Encourage your child to talk about some of the math ideas shown on page 411.

Look for the 🏠 at the bottom of each skills lesson. The suggestions on the page give you an opportunity to improve your child's understanding of math. You may want to have pennies and other small countables available for your child to use throughout this chapter.

Home Activity

Fish Facts

Try this activity with your child. Draw two nets (see diagram below). Draw a total of 13 fish on the two nets. Ask your child to tell the addition sentence shown below. Cross out 4 fish on one of the nets, then ask her/him to tell the subtraction sentence. After each lesson, adjust the total number of fish and the number on the sign to correspond with addition and subtraction facts 13 through 18.

Home Reading Connection

At the Beach

Today at the beach
We made up a game.

With thick short sticks
We scribbled our names.

Twelve letters in all
(Samantha and Josh,)

and then came the wave,
and we saw the sea wash

our names from the beach
but left in its trail

six smooth white shells
to put in our pails.

Rebecca Kai Dotlich

MATH CONNECTIONS
Missing Numbers

Name _____

Judy wrote number sentences in the sand. Then she hid one part with shells.
Find the missing numbers.

Part + Part = Whole

8 = 5 + ⬚(3)

10 = 9 + ⬚

⬚ = 6 + 6

⬚ = 5 + 4

4 = 5 − ⬚

7 = 10 − ⬚

⬚ = 8 − 2

⬚ = 7 − 4

⬚ + 7 = 9

⬚ + 3 = 6

10 − ⬚ = 5

11 − ⬚ = 9

⬚ − 1 = 7

⬚ − 2 = 10

Which shells hide the number in all? Color them 🖍 red.

Which shells hide a part? Color them 🖍 blue.

Whole − Part = Part

You can put this in your Math Portfolio.

four hundred thirteen 413

CROSS-CURRICULAR CONNECTIONS
Art and Space Figures

Name _____

Tess built this winning sand castle. Tally and graph the containers she used.

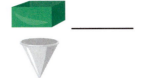

  ____

  ____

____  ____

____  ____

rectangular prism

cone

cube

pyramid

cylinder

pail

Castle Containers

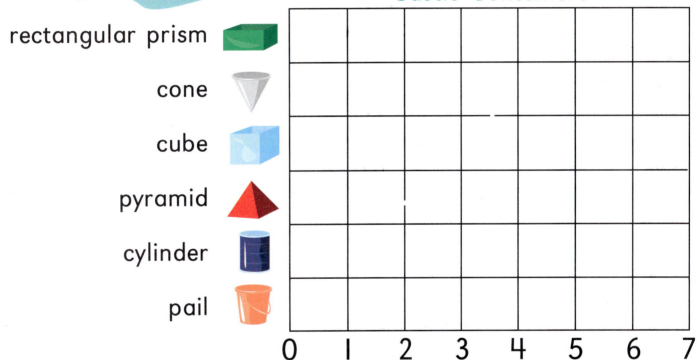

Use the beach ball.
Color to make each number sentence correct.

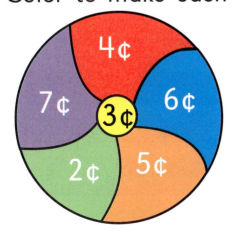

🟥 + 🟥 + ☐ = 10¢

🟪 + 🟨 + ☐ = 14¢

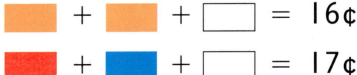

🟧 + 🟧 + ☐ = 16¢

🟥 + 🟦 + ☐ = 17¢

414 four hundred fourteen

You can put this in your Math Portfolio. **PORTFOLIO**

Name _____

Sums of 13 and 14

6 + 6 = 12
So 7 + 7 = 14.

7 + 7 is a doubles fact.

6 + 7 and
7 + 6 are 1 less
than 7 + 7.

```
  6       7
 +7      +6
 ──      ──
 13      13
```

Add. Write the related fact.

1.
```
  8
 +5
 ──
 13
```
  +8
 ──
 13

2.
```
  4
 +9
 ──
```

3.
```
  6
 +8
 ──
```

4.
```
  5
 +9
 ──
```

 ✓ sums of 14 in 1–4.

11-1 Ask your child to model the 6 sums of 13 and the 5 sums of 14 shown above.

four hundred fifteen **415**

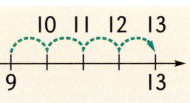

"I count on from the greater addend."

4¢ + 9¢ = 13¢

Why did Rita start with 9? Why did Rita stop at 13?

Find the sum.

1.
5¢	8¢	6¢	3	4	7
+7¢	+5¢	+7¢	+8	+9	+4
12¢					

2.
7¢	9¢	9¢	4	5	9
+6¢	+5¢	+4¢	+8	+8	+3

3.
6¢	6¢	7¢	3	8	5
+8¢	+6¢	+7¢	+9	+6	+9

 Ring sums between 12 and 14 in 1–3.

 4. Lori has 9¢. She finds 5¢ more. How much does she have now? ____

5. Josh lost 8¢. Then he lost a nickel. Has he lost more or less than 14¢? ____

 6. Complete the pattern.

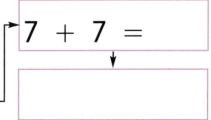

5 + 7 =
6 + 7 =
7 + 7 =

416 four hundred sixteen

Name _____

Subtract from 13 and 14

Subtract across or down.   14
14 − 7 = 7   − 7
whole part part   ———
                            7

Start by taking away the last part.

Subtract. Write the related fact.

1. 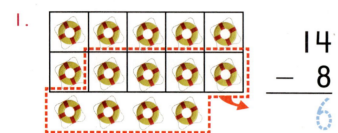   14
                          − 8
                          ———
                           6

   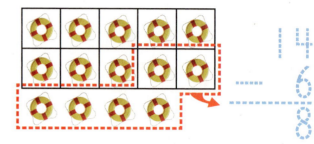   14
                          − 6
                          ———
                           8

2. 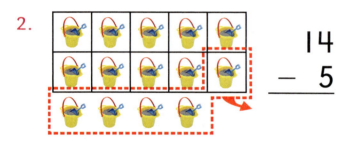   14
                          − 5
                          ———

   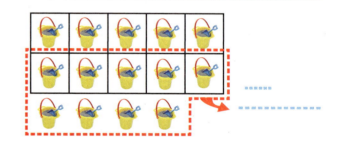   ___
                          − ___
                          ———

3. 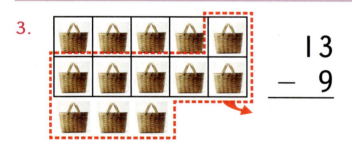   13
                          − 9
                          ———

   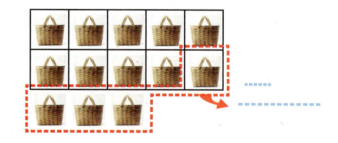   ___
                          − ___
                          ———

4. 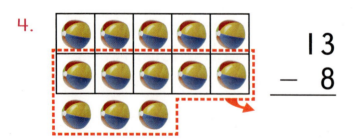   13
                          − 8
                          ———

   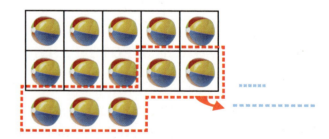   ___
                          − ___
                          ———

5. 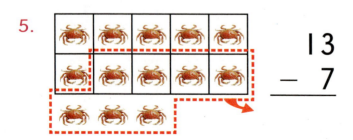 13
                          − 7
                          ———

   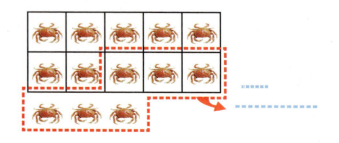 ___
                          − ___
                          ———

11-2 Prompt your child to explain how she/he takes away from 13 or 14.

"I can use doubles to find the difference."

"I can use related subtraction facts, too!"

14 − 7 = 7
So 14 − 6 = 8.

13 − 9 = 4
So 13 − 4 = 9.

Find the difference.

1.
13	12	13	11¢	13¢	14¢
− 5	− 4	− 6	− 3¢	− 4¢	− 9¢
8					

2.
13	12	13	11¢	13¢	14¢
− 7	− 7	− 8	− 7¢	− 9¢	− 6¢

3.
14	12	12	14¢	12¢	14¢
− 8	− 3	− 8	− 7¢	− 5¢	− 5¢

 Which strategy did you use for each fact in 1–3?

4. Subtract 8 from
8	9	10	11	12	13	?
0						?

5. Subtract 6 from
6	8	10	12	14	16	?

418 four hundred eighteen

Name _____

Sums of 15 and 16

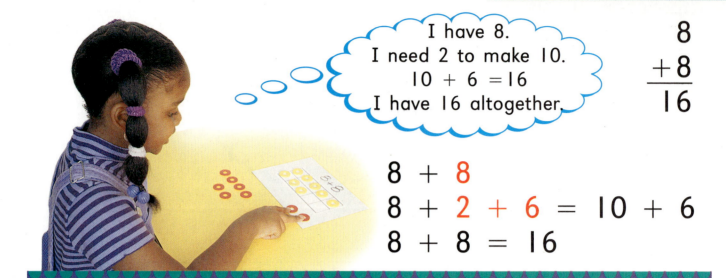

I have 8.
I need 2 to make 10.
10 + 6 = 16
I have 16 altogether.

$$\begin{array}{r} 8 \\ +8 \\ \hline 16 \end{array}$$

8 + **8**
8 + **2** + **6** = 10 + 6
8 + 8 = 16

Draw the missing addend. Find the sum.

1. 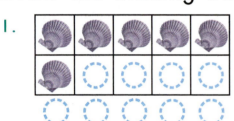 $\begin{array}{r} 6 \\ +9 \\ \hline 15 \end{array}$

2. $\begin{array}{r} 9 \\ +6 \\ \hline \end{array}$

3. $\begin{array}{r} 7 \\ +8 \\ \hline \end{array}$

4. 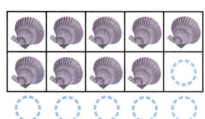 $\begin{array}{r} 8 \\ +7 \\ \hline \end{array}$

5. $\begin{array}{r} 7 \\ +9 \\ \hline \end{array}$

6. 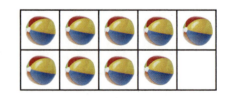 $\begin{array}{r} 9 \\ +7 \\ \hline \end{array}$

 TALK IT OVER How did you make 10 to find each sum in 1–6?

11-3 Give your child 15 or 16 countables and have him/her explain how to add by making ten.

four hundred nineteen **419**

9 + 6 = 15
6 + 9 = 15

This number line shows two **related facts**.

Add.

1. 9   8   6   4
  +4   +8   +8   +9
  13

2. 5   9   9   8   6
  +8   +6   +5   +5   +7

3. 8   6   9   8   7   5   6
  +7   +6   +3   +6   +8   +7   +9

4. 4   7   3   9   7   5   7
  +8   +9   +9   +7   +6   +9   +7

 ✓ the related facts in each row in 1–4.

 5. Nina saved 8 pennies. Then she saved 7 more. Can she trade her pennies for 3 nickels? Why? _____

420 four hundred twenty

Name _____

Subtract from 15 and 16

Look for a pattern when you subtract 9.

12 − 9 = 3
13 − 9 = 4
14 − 9 = 5
15 − 9 = ?

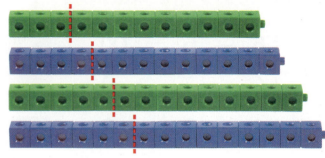

15 − 9 = 6

Subtract. Write the related fact.

1. 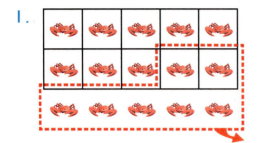 15 − 7 = 8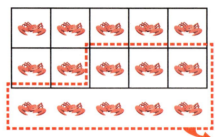

2. 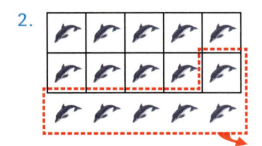 15 − 6   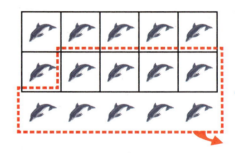 _____

3. 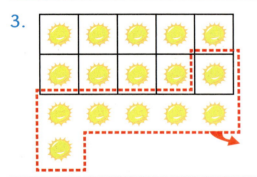 16 − 7   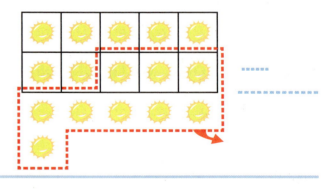 _____

4. 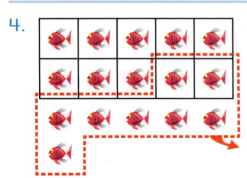 16 − 8   Does 4 have a related fact? Explain.

Tell your child to use countables to show subtraction patterns with 9 for facts up to 18.

four hundred twenty-one 421

Use patterns to subtract.

Part left   Part taken away

16 − 6 = 10
16 − 7 = 9
16 − 8 = 8

 Describe the pattern in the subtraction sentences. What are the next 2 subtraction sentences?

Find the difference.

1. 16   15   13   15   14   12
   − 7   − 7   − 9   − 8   − 7   − 3
    9

2. 14   13   16   14   12   11
   − 9   − 5   − 8   − 8   − 9   − 6

3. 15   16   13   15   13   14
   − 6   − 9   − 8   − 9   − 6   − 5

 ✓ odd differences in 1−3.

 Write the missing whole.

4. ___ − 5 = 10   5. ___ − 9 = 7
   ___ − 6 = 9   ___ − 8 = 7
   ___ − 7 = 8   ___ − 7 = 7

Whole − Part = Part

422 four hundred twenty-two

Name _____

Facts of 17 and 18

8 + 8 = 16
So 8 + 9 = 17.

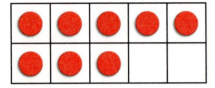

10 + 10 = 20
So 9 + 9 = 18.

 TALK IT OVER What is the related addition fact for 8 + 9? Name another way to find 9 + 9.

Write the second addend. Add.

1.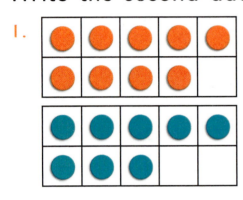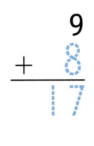
$\begin{array}{r} 9 \\ + 8 \\ \hline 17 \end{array}$

2.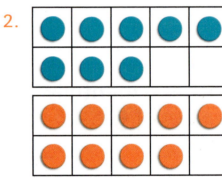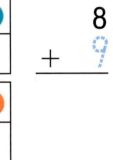
$\begin{array}{r} 8 \\ + 9 \\ \hline \end{array}$

3. 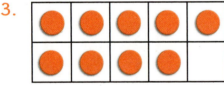 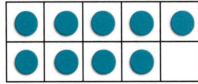   9 _____

Find the sum.

4. 8 + 6 = 14   8 + 9 = ___   6 + 9 = ___

5. 7 + 8 = ___   9 + 8 = ___   7 + 9 = ___

6. 7 + 7 = ___   9 + 9 = ___   8 + 8 = ___

 ✔ even sums in 4—6.

11-5 Direct your child to draw two parts that make up a whole of 17 and then two parts that make up a whole of 18.

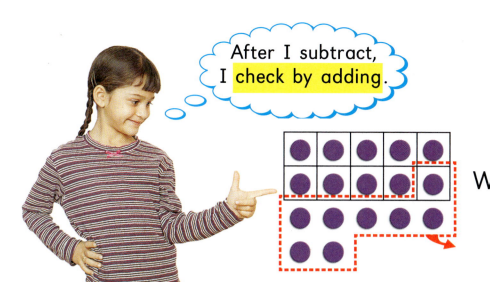

Find the difference. Check by adding.

1. 17 − 9 = 8   + 9 =
2. 18 − 9 =   + 9 =

Subtract. Check mentally.

3. 17   15   18   13   15   13
   − 8 − 9 − 9 − 7 − 6 − 7
    9

4. 13   14   14   17   16   15
   − 5 − 6 − 7 − 9 − 8 − 8

5. 14   16   16   14   13   14
   − 5 − 9 − 7 − 8 − 4 − 9

 6. What pattern do you see in the differences in 3–5? Write a pattern of your own.

424   four hundred twenty-four

Name _____

Fact Families

Yuri has these 3 cards.   8 13 5

He makes a **fact family**.

8 + 5 = 13   5 + 8 = 13
13 − 5 = 8   13 − 8 = 5

Talk it Over: How does knowing related facts help you find a fact family?

Write the fact family.

1. 5 14 9
   _5_ + _9_ = ___   ___ + ___ = ___
   ___ − ___ = ___   ___ − ___ = ___

2. 13 6 7
   ___ + ___ = ___   ___ + ___ = ___
   ___ − ___ = ___   ___ − ___ = ___

3. 17 8 9
   ___ + ___ = ___   ___ + ___ = ___
   ___ − ___ = ___   ___ − ___ = ___

4. 16 8
   ___ + ___ = ___   ___ − ___ = ___

CHALLENGE
Write the fact family.
✗ the card that does not belong.

5. 9 14 13 4
   ___ + ___ = ___   ___ + ___ = ___
   ___ − ___ = ___   ___ − ___ = ___

11-6 Say numbers in a fact family, such as 8, 9 and 17, and have your child write the fact family.

Name _____

Ring Yes or No.

1. A 🥛 holds about 1 cup. (Yes)   No

2. A 🚢 is shorter than 1 foot.   Yes   No

3. A 👡 is about 1 inch long.   Yes   No

4. A 🫗 holds more than 1 quart.   Yes   No

5. A 🍶 holds about 1 liter.   Yes   No

6. The 🧒 weighs more than 1 kilogram.   Yes   No

7. A 🪣 is about 1 foot long.   Yes   No

8. A 🐬 is lighter than 1 pound.   Yes   No

PROBLEM SOLVING

9. What is the distance around in shells?   ____ shells

10. Ring the best tool to find the temperature.

426   four hundred twenty-six

PROBLEM-SOLVING STRATEGY
Extra Information

Some problems have extra information.

1. **Read** Ryan found 8 .
   Niki found 7 .
   ~~4 were big.~~
   How many did they find in all?

   Draw a line through the sentence that you do not need.

   Think The 3rd sentence is extra information.

   Write __8__ ⊕ __7__ = ___

   Check Use models to check. ___ in all

2. **Read** Lauren saw 14 🐦 in the sky.
   Jim saw 9 🐦 on the beach.
   2 🐦 were eating.
   How many more 🐦 did Lauren see than Jim?

   Think The __3rd__ sentence is extra information.

   Write ___ ◯ ___ = ___

   Check Use models to check. She saw ___ more 🐦.

3. Raul made 8 🏰. Heather made 6 🏰.
   Raul worked for 1 hour.
   How many 🏰 were there in all?

   The ____ sentence is extra information.

   ___ ◯ ___ = ___   They made ___ 🏰 in all.

11-8 Prompt your child to explain how he/she knows if a problem has a fact that is not needed.

four hundred twenty-nine **429**

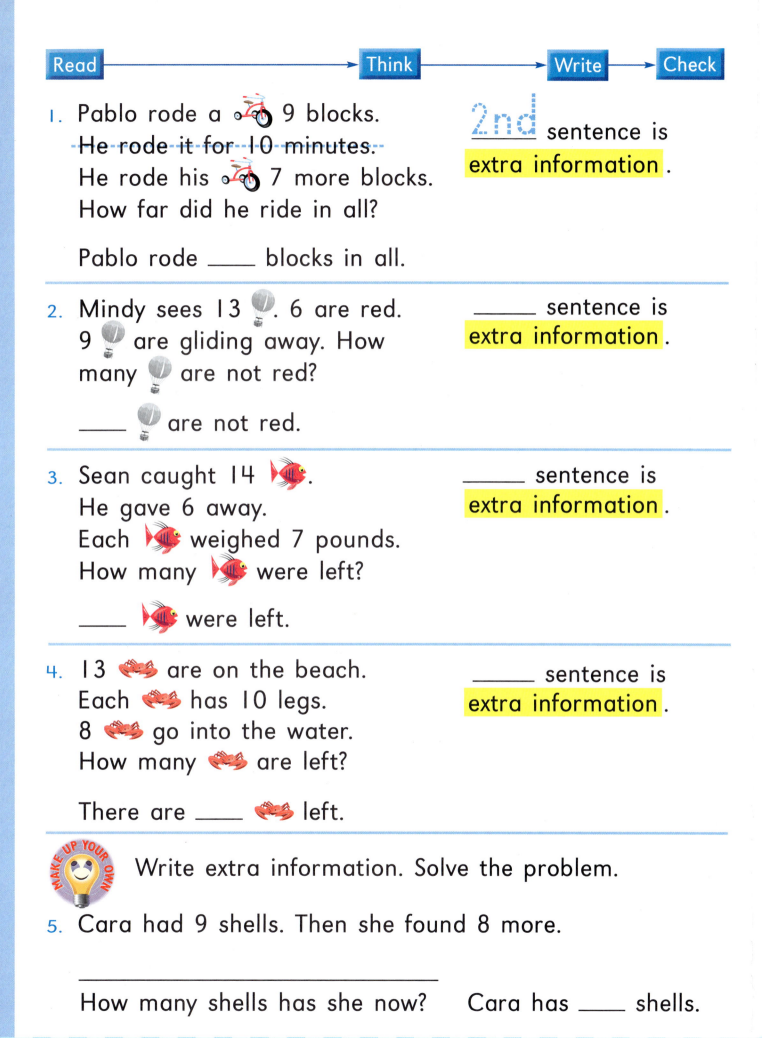

Read → Think → Write → Check

1. Pablo rode a 🚲 9 blocks.
   ~~He rode it for 10 minutes.~~
   He rode his 🚲 7 more blocks.
   How far did he ride in all?

   Pablo rode ____ blocks in all.

   __2nd__ sentence is extra information.

2. Mindy sees 13 🎈. 6 are red.
   9 🎈 are gliding away. How many 🎈 are not red?

   ____ 🎈 are not red.

   ____ sentence is extra information.

3. Sean caught 14 🐟.
   He gave 6 away.
   Each 🐟 weighed 7 pounds.
   How many 🐟 were left?

   ____ 🐟 were left.

   ____ sentence is extra information.

4. 13 🦀 are on the beach.
   Each 🦀 has 10 legs.
   8 🦀 go into the water.
   How many 🦀 are left?

   There are ____ 🦀 left.

   ____ sentence is extra information.

Write extra information. Solve the problem.

5. Cara had 9 shells. Then she found 8 more.
   _____
   How many shells has she now?   Cara has ____ shells.

Performance Assessment

Name _____

1. Complete each fact to show the strategy. Explain your thinking.

   • Use doubles.   8 + ___ = ___

   • Count on from the
     greater addend.   9 + ___ = ___

   • Check by adding.   13 − 7 = ___

                              ___ + ___ = ___

   • Make 10.   ___ + 6 + ___ = ___

2. Use the strategies to find each difference. Write the facts you used.

Use doubles	Use related facts
16¢ − 7¢ ___	13¢ − 7¢ ___

PORTFOLIO

Use a separate sheet of paper. Choose 1 project.

3. Predict the sums with more facts. Ring your prediction. Make a list to check.

   sums of 15
   or
   sums of 16

4. Use 3 . Show the sum of a doubles fact plus 1 on each. Are the sums odd or even?

   5 + 5 + 1 = ?

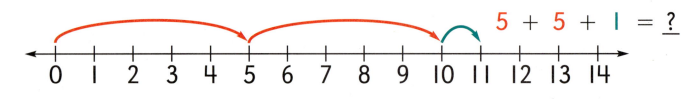

Check Your Mastery

Name _____

Add or subtract. Watch for + and −.

1. 8   9   8   7¢   6¢   5¢
   +7   +5   +9   +7¢   +7¢   +8¢
   ___ ___ ___ ____ ____ ____

2. 13   17   14   16¢   14¢   15¢
   − 4 − 8 − 5 − 9¢ − 9¢ − 7¢
   ___ ___ ___ ____ ____ ____

3. 15   6   14   13   7   9
   − 9 +8 − 6 − 7 +8   −6
   ___ ___ ___ ___ ___ ___

Add. Ring the group you use.

4. 9 + 1 + 4 = ___   5. 4 + 4 + 9 = ___

6. 7 + 2 + 8 = ___   7. 8 + 3 + 5 = ___

PROBLEM SOLVING Use a strategy you have learned.

8. Cleo sees 7 .
   Dan sees 8 🚤.
   4 of them are blue.
   How many 🚤 are there in all?

9. Lois fills 15 🪣 and George fills 8.
   How many more 🪣 does Lois fill than George?

   There are ___ in all.

   Lois fills ___ more .

436 four hundred thirty-six

Cumulative Test II
Chapters 1–11

Name _____

Mark the ○ for your answer.

Listening Section

A

B 1st
○ 12 ○ 13 ○ 14

1. What is the amount?
○ 16¢
○ 26¢
○ 41¢
○ 46¢

2. What time is it?
○ 6:00
○ 6:30
○ 11:30
○ 12:30

3. How many centimeters?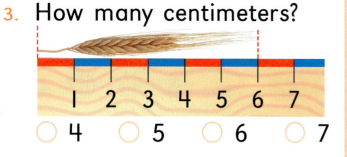
○ 4 ○ 5 ○ 6 ○ 7

4. Which shows halves?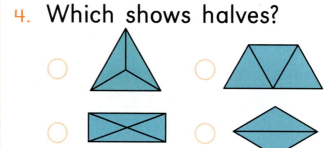

5. What is the same shape as this ◁ ?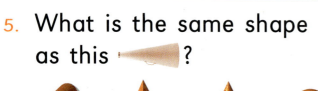

6. Which holds less than 1 pint?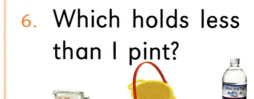

7. 6
   +7

○ 13
○ 14
○ 15
○ 16

8. 14
   − 9
○ 3
○ 4
○ 5
○ 15

9. 6
    2
   +8
○ 15
○ 16
○ 17
○ 18

10. Measure in inches.
○ 2 ○ 4
○ 3 ○ 5

four hundred thirty-seven **437**

Mark the ◯ for your answer.

11. 20, 30, 40, 50, ____
    ◯ 90   ◯ 70   ◯ 60   ◯ 40

12. 8 + 7 and 9 + 6 are names for:
    ◯ 17   ◯ 15   ◯ 13   ◯ 11

13. What is between 39 and 41?
    ◯ 37   ◯ 40   ◯ 42   ◯ 38

14. 4 tens 2 ones + 3 ones
    ◯ 75   ◯ 72   ◯ 45   ◯ 42

15. 63
    +14
    ◯ 51 ◯ 57 ◯ 67 ◯ 77

16. 58
    −16
    ◯ 42 ◯ 44 ◯ 52 ◯ 74

17. 53¢
    +32¢
    ◯ 21¢ ◯ 25¢ ◯ 55¢ ◯ 85¢

18. Which is greater than 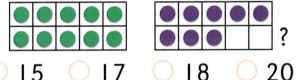 ?
    ◯ 15   ◯ 17   ◯ 18   ◯ 20

19. 6 ● = ____ groups of 2.

    ◯ 2   ◯ 3   ◯ 4   ◯ 6

20. What does not belong to the same fact family?
    ◯ 8 + 9 = 17
    ◯ 17 − 9 = 8
    ◯ 9 + 9 = 18
    ◯ 17 − 8 = 9

21. What comes next?
    8 − 4, 10 − 5, 12 − 6,
    ◯ 12 − 7   ◯ 14 − 7
    ◯ 16 − 8   ◯ 18 − 9

Weight in Pounds

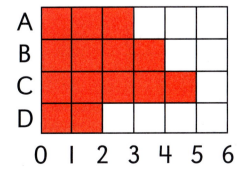

22. A is 2 pounds less than ____.
    ◯ D   ◯ C   ◯ A   ◯ B

23. B, C, and D weigh ____ pounds altogether.
    ◯ 14   ◯ 13   ◯ 12   ◯ 11

Moving On in Math

12

CRITICAL THINKING

Nikki had between 10 and 20 stars. Put in sixes, there were none left over; when put in sevens, there were 4 left over. How many stars were there?

Math Alive at Home

Dear Family,

Today your child began Chapter 12. As she/he uses algebraic reasoning and extends her/his knowledge to 3-digit numbers, you may want to read the poem below, which was read in class. Encourage your child to talk about some of the math ideas shown on page 439.

Look for the 🏠 at the bottom of each lesson. You can use the suggestion to improve your child's understanding of the skills she/he learned.

For more information about Chapter 12, visit the Family Information Center at **www.sadlier-oxford.com**

Home Reading Connection

Stars

The stars are too many to count.
The stars make sixes and sevens.
The stars tell nothing—and everything.
The stars look scattered.
Stars are so far away they never speak when spoken to.

Carl Sandburg

Home Activity

Star Track

Have your child name a sum and a difference for a number in a star track like the one shown. After your child says both number sentences, have her/him color the star. You can change the numbers to reflect different skills beginning on page 447.

Name _____

MATH CONNECTIONS
Fractions and Probability

Complete the table.

1. __4__ out of __8__ are ⭐ (yellow).
2. ____ out of ____ are ⭐ (blue).
3. ____ out of ____ are ⭐ (red).

$\frac{4}{8}$

Without looking, Tia picks a star.
Write **always**, **most**, **never**, or **least**.

4. Tia is _____ likely to pick ⭐ (yellow).

5. Tia will _____ pick ⭐ (green).

6. Tia is _____ likely to pick ⭐ (blue).

7. She will _____ pick a star.

 What can Tia do to make the chance of picking each color the same?

8. Color to make the chance of picking ⭐, ⭐, and ⭐ equal. Then write each fraction.

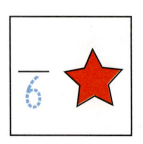 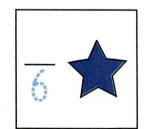 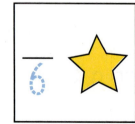

You can put this in your Math Portfolio.

four hundred forty-one **441**

CROSS-CURRICULAR CONNECTIONS
Science

Name _____

Stars are hot balls of gas.
Stars differ in temperature.
Color the stars.

Temperature	Color
warm	★ (red)
hot	★ (yellow)
hotter	☆ (white)
hottest	★ (blue)

1. red — less than thirty
2. yellow — thirty to forty-nine
3. white — fifty to seventy
4. blue — greater than seventy

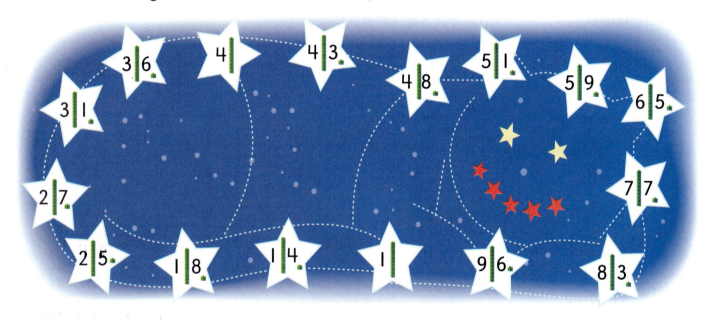

5. Start at 10. Connect the stars in order. Then name your constellation. _____

Describe your constellation.

6. It has _____ stars altogether.

7. It has _____ more ★ than ★.

8. It has _____ less ★ than ★.

Draw 3 blue stars for a tail.

9. Now it has _____ stars in all.

You can put this in your Math Portfolio.

442 four hundred forty-two

Name _____

Find the Rule

The dominoes follow the same rule.

3 + 2 = 5   1 + 2 = 3   4 + 2 = 6

Rule: Dots on the left plus 2 equal dots on the right.

Use the rule above. Write the number sentence.

1.  ____ + 2 = ____

Write the number sentence for each.
Draw another domino and number sentence for the same rule.

2.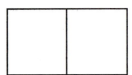

____ − 3 = 1   ____ = 3   ____

3.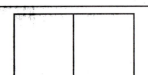

____ = 6   ____ = 2   ____

4.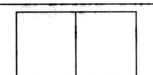

____ = 6   ____ = 2   ____

 Is there more than 1 rule for 3?
Tell the rule for 4.

 5. Use a separate sheet of paper.
Draw 3 dominoes for a subtraction rule.

Ask your child to draw 3 dominoes for the rule: dots on the left minus 2.

four hundred forty-three 443

Number Sentence Balance

Name _____

Find the missing number.

15 − 9 = 3 + ▢

> 15 − 9 = 6
> 3 + ? = 6

15 − 9 = 3 + 3
 ↓   ↓
 6   =   6

▢ = 3

Fill in the ▢. Solve to check.

1. 17 − 8 = [5] + 4
   9 = 9

2. 8 + 5 = ▢ + 3
   ___ = ___

3. ▢ + 8 = 7 + 9
   ___ = ___

4. ▢ − 10 = 7 + 3
   ___ = ___

5. 25 − ▢ = 7 + 8
   ___ = ___

6. 28 − ▢ = 4 + 4
   ___ = ___

7. 38 − 20 = ▢ + 9
   ___ = ___

8. 19 − 6 = ▢ + 7
   ___ = ___

9. 9 + 4 = 10 + ▢
   ___ = ___

10. 37 − 23 = 8 + ▢
    ___ = ___

FINDING TOGETHER

11. Write 3 different number sentences for each balance.

    △ + 8 = 25 − ▧   4 + ▢ = ▱ − 5

444 four hundred forty-four

Name _____

Missing Operations

You can use **guess and test** to find the missing signs.

17 ◯ 10 ◯ 2 = 9

Add or subtract left to right.

Try − and −.   17 ⊖ 10 ⊖ 2 = 5
Try + and −.   17 ⊕ 10 ⊖ 2 = 25
Try − and +.   17 ⊖ 10 ⊕ 2 = 9

So the missing signs are − and +.

Talk it Over: Why would + and + not be a good guess?

Fill in the missing signs.

1. 8 ◯ 5 ◯ 3 = 10

2. 12 ◯ 3 ◯ 8 = 1

3. 6 ◯ 9 ◯ 10 = 25

4. 8 ◯ 7 ◯ 6 = 9

5. 12 ◯ 5 ◯ 4 = 3

6. 5 ◯ 9 ◯ 8 = 6

7. 5 ◯ 8 ◯ 10 = 23

Guess and test.

Ask your child to explain how he/she would find the missing signs in 10 ◯ 6 ◯ 9 = 13.

four hundred forty-five **445**

Missing Numbers

Name _____

When 🟦 = 1, what does 🔺 equal?

🟦 + 🔺 = 10

1 + 🔺 = 10

Think: 1 + 9 = 10

So 🔺 = 9.

Write 1 for the 🟦. Then find the missing number.

Find the missing number.

1. 🟧 = 6   🔶 − 🟧 = 9   15 − 6 = 9
   🔶 = ?   🔶 − 6 = 9   🔶 = 15

2. 🟡 = 28   🟡 − 🟦 = 18   ___
   🟦 = ?   ___ − 🟦 = ___   🟦 = ___

3. 🔺 = 7   🟧 + 🔺 = 16   ___
   🟧 = ?   🟧 + ___ = ___   🟧 = ___

4. 🟨 = 15   🟨 − 🔴 = 8   ___
   🔴 = ?   ___ − 🔴 = ___   🔴 = ___

5. 🔷 = 40   🔷 + 🟧 + 1 = 61   ___
   🟧 = ?   ___ + 🟧 + ___ = 61   🟧 = ___

446 four hundred forty-six

Have your child explain how he/she found the missing number.

12-4

Regrouping Money

Trade 10 ¢ for 1 ⊙ when you regroup to add.

Add pennies. Regroup. Add dimes.

dimes	pennies
¹3	2
+ 2	8
6	0

Write the cent sign.

```
  32¢
+ 28¢
  60¢
```

Find the sum.

1.
¹57¢ + 24¢ = 81¢
48¢ + 11¢ = 59¢
36¢ + 34¢
13¢ + 58¢
27¢ + 47¢

2.
67¢ + 26¢
33¢ + 59¢
78¢ + 12¢
45¢ + 17¢
63¢ + 24¢

3.
24¢ + 8¢
19¢ + 1¢
54¢ + 4¢
81¢ + 9¢
68¢ + 4¢

4.
32¢ + 5¢
46¢ + 8¢
29¢ + 2¢
65¢ + 9¢
88¢ + 3¢

5.
49¢ + 9¢
79¢ + 6¢
45¢ + 45¢
30¢ + 15¢
87¢ + 3¢

12-5 Have your child explain how to regroup money in addition and subtraction.

four hundred forty-seven

Trade 1 🪙 for 10 🟠 to regroup in subtraction.

Regroup to have enough pennies. 3 dimes 1 penny = 2 dimes 11 pennies.

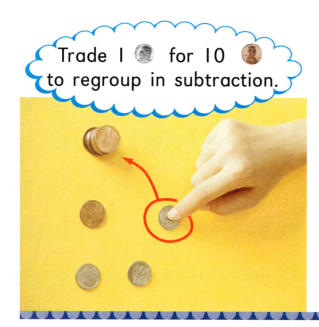

dimes	pennies
²3̸	¹¹1̸
−1	6
1	5

² ¹¹
3̸1̸¢
−16¢
15¢

Find the difference. Regroup where needed.

1. ⁶¹⁴7̸4̸¢   ⁴¹⁰5̸0̸¢   36¢   65¢   72¢
    −18¢   −27¢   −16¢ −29¢ −16¢
    56¢   23¢

2. 84¢   23¢   91¢   80¢   40¢
    − 1¢ − 6¢ − 9¢ − 5¢ − 1¢

3. 67¢   21¢   50¢   33¢   99¢
    −39¢ − 3¢ −19¢ − 4¢ −87¢

4. 46¢   60¢   35¢   96¢   30¢
    − 8¢ −43¢ −33¢ −57¢ − 4¢

5. Ryan has 60¢. How much money does he need to buy both toys? _____

35¢

49¢

Name _____

Add Ones: Regroup

57 + 8 = ?

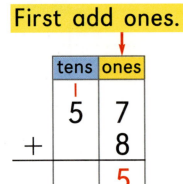

15 ones equals
1 ten 5 ones.

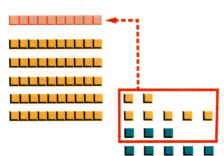

Find the sum. Use ▭▭▭▭ and ▪ to model.

1.
tens	ones
2	5
+	5
3	0

tens	ones

tens	ones
3	4
+	9

tens	ones
1	1
+	9

2.
tens	ones
4	7
+	6

tens	ones
6	6
+	5

tens	ones
2	4
+	8

tens	ones
8	9
+	2

3.
tens	ones
5	8
+	6

tens	ones
1	5
+	7

tens	ones
3	7
+	9

tens	ones
6	9
+	8

TALK IT OVER Why did you need to regroup in 1–3?

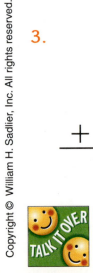

Ask your child to redo an addition exercise and tell why and how he/she regrouped.

four hundred forty-nine **449**

Write each addend. Add.

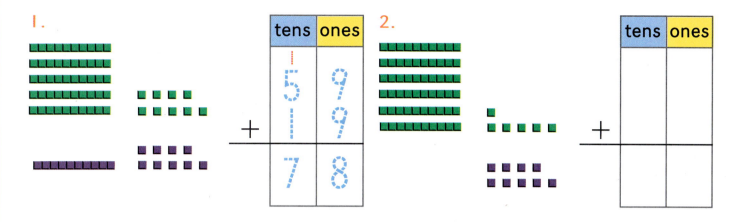

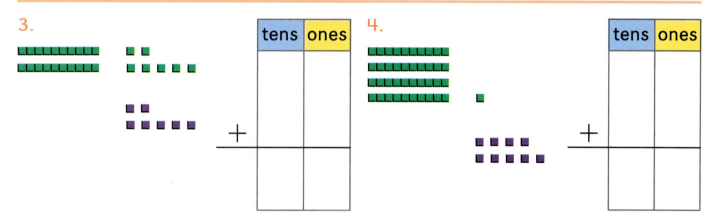

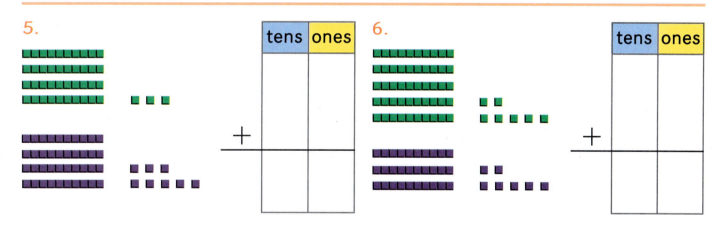

 Use ▬▬▬ and ▪.

7. Ricardo has 43 gold stars.
Rosa gave him 7 more stars.
How many stars does Ricardo have now?

Name _____

Subtract Ones: Regroup

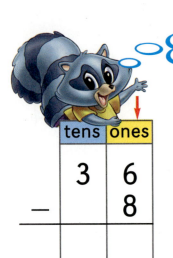

You need more ones to subtract 36 − 8.

Regroup 3 tens 6 ones as 2 tens 16 ones.

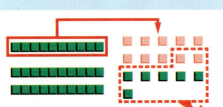

Now subtract. Begin with ones.

tens	ones
³2̶	¹⁶6̶
	8
2	8

tens	ones
3	6
	8

Subtract. Ring to take away.

1.
tens	ones
³4̶	¹⁰0̶
	7
3	3

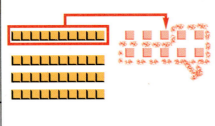

2.
tens	ones
6	5
	6

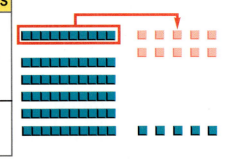

3.
tens	ones
8	2
	5

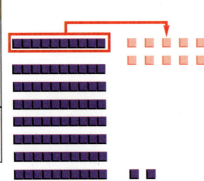

4.
tens	ones
7	1
	4

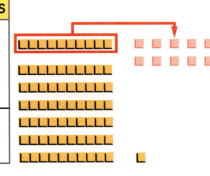

In 5 tell how you would regroup to subtract each.
How many tens will be left?

5.
tens	ones
2	0
	8

tens	ones
2	0
	4

tens	ones
2	0
	2

Have your child explain how he/she would regroup to find 20 − 3.

four hundred fifty-one 451

Find the difference. Use ▭▭▭ and ▪.

1.

tens	ones
²3̸	¹²2̸
	8
2	4

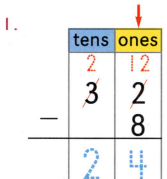

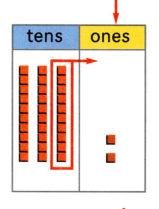

tens	ones
3	2
	7

tens	ones
2	4
	8

2.

tens	ones
2	5
	8

tens	ones
3	3
	9

tens	ones
2	1
	6

tens	ones
5	0
	3

3.

tens	ones
3	6
	8

tens	ones
4	0
	9

tens	ones
2	6
	9

tens	ones
3	6
	7

4.

tens	ones
4	7
	8

tens	ones
2	2
	4

tens	ones
4	5
	7

tens	ones
4	1
	2

5. Sue buys both toys. She pays with 1 🪙. How much money will she have left?

Sue will have ____ left.

3¢   9¢

452   four hundred fifty-two

Name _____

Three Addends: No Regrouping

Find the sum of 23, 11, and 14.

First add ones.

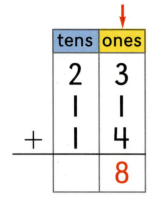

tens	ones
2	3
1	1
+ 1	4
	8

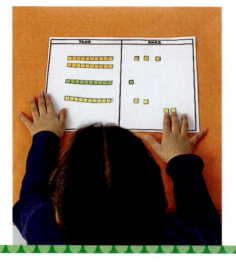

Then add tens.

tens	ones
2	3
1	1
+ 1	4
4	8

Find the sum.

1. 11   14   13   12   32
    32   21   65   32   20
   +24   +63   +11   +35   +35
   ────
    67

2. 25   12   16   13   21
    31   41   11   11   37
   +23   +23   +62   +74   +40

3. 42   21   11   13   23
    20   53   28   54   12
   +22   +13   +30   +12   +63

In 1–3 ✔ odd sums.

When did you add up? When did you add down?
When did you use doubles?

Ask your child to show you examples of when he/she used doubles to add.

four hundred fifty-three 453

Write the missing addends.

1. 22 + ___ + ___ = 75
2. 41 + ___ + ___ = 96
3. 12 + ___ + ___ = 59
4. 22 + ___ + ___ = 48
5. 14 + ___ + ___ = 77
6. 33 + ___ + ___ = 88
7. 33 + ___ + ___ = 86
8. 12 + ___ + ___ = 67
9. 22 + ___ + ___ = 69
10. 21 + ___ + ___ = 68

11. What is the sum of the numbers in the circle? ___

12. What is the sum of the numbers not in the rectangle? ___

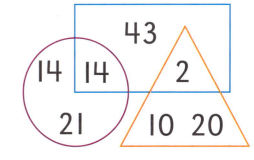

 Look for tens.

13. (6 + 4) + 27 = 37
14. 5 + 45 + 5 = ___
15. 16 + 8 + 2 = ___
16. 3 + 32 + 7 = ___
17. 28 + 1 + 9 = ___
18. 2 + 51 + 8 = ___

Name _____

Hundreds, Tens, Ones

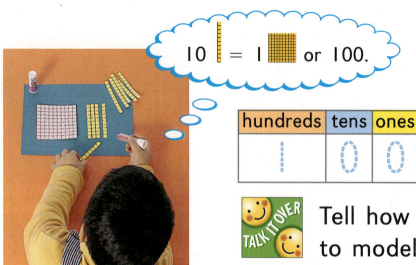

10 | = 1 ▦ or 100.

hundreds	tens	ones
1	0	0

hundreds	tens	ones
2	0	0

two hundred

TALK IT OVER Tell how to model 400. Tell how to model this page number.

Write the place value and number word.

1.

   _2_ hundreds _0_ tens _3_ ones

hundreds	tens	ones
2	0	3

   two hundred three

2. ___ hundreds ___ ten ___ ones

hundreds	tens	ones

   three hundred twelve

3.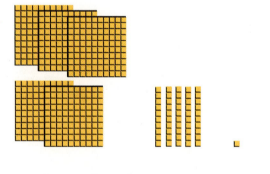

   ___ hundreds ___ tens ___ one

hundreds	tens	ones

   five hundred fifty-one

4.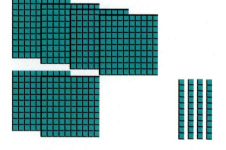

   ___ hundreds ___ tens ___ ones

hundreds	tens	ones

   six hundred forty

12-9 Ask your child to write a 3-digit number and tell how many hundreds, tens, and ones it has.

Color to show how many hundreds, tens, and ones.

three hundred eight

1.
hundreds	tens	ones
3	0	8

two hundred forty-five

2.
hundreds	tens	ones
2	4	5

two hundred ten

3.
hundreds	tens	ones
2	1	0

 Share Your Thinking

Which number in 1–3 is greatest?

Which is the least? How do you know?

 Problem Solving Use 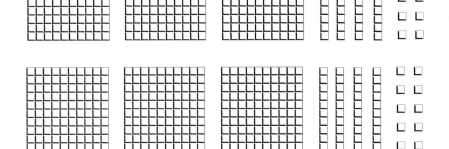, and.

4. There are 365 days in 1 year. A  has been in space 10 days less than 1 year and a  100 days more than 1 year. How long has each been in space?

 Critical Thinking

Ring the group that shows the red number.

5. 3**2**1

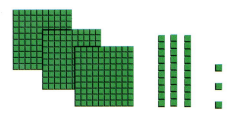

6. 2**3**1

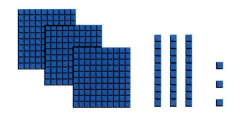

456 four hundred fifty-six

Name _____

Add and Subtract Three-Digit Numbers

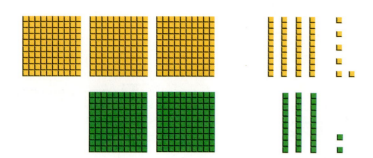

Add ones. | Next add tens. | Then add hundreds.

h	t	o
3	4	6
+2	3	2
		8

h	t	o
3	4	6
+2	3	2
	7	8

h	t	o
3	4	6
+2	3	2
5	7	8

346 + 232 = ?

346 + 232 = 578

Find the sum. Use 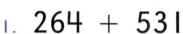.

1. 264 + 531

h	t	o
2	6	4
+5	3	1
7	9	5

2. 472 + 517

h	t	o
+		

3. 536 + 432

h	t	o
+		

4. 128   263   364   612   347
   +370   +212   +304   +244   +642

5. 551   401   225   535   139
   +303   +337   +464   +261   +450

12-10 Ask your child to find the sum of 739 + 250.

four hundred fifty-seven **457**

218 − 103 = ?

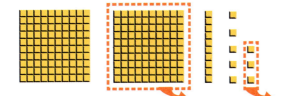

First subtract ones. **Next subtract tens.** **Then subtract hundreds.**

h	t	o
2	1	8
− 1	0	3
		5

h	t	o
2	1	8
− 1	0	3
	1	5

h	t	o
2	1	8
− 1	0	3
1	1	5

218 − 103 = 115

Subtract. Use ▦, ▭, ▪.

1. 576 − 341

h	t	o
5	7	6
−3	4	1
2	3	5

2. 374 − 152

3. 253 − 130

4.
784
−251

564
−432

397
−145

479
−336

698
−434

Find the missing number.

5.
853
−3☐0
543

679
−☐17
462

984
−2☐4
720

475
−20☐
272

916
−☐03
513

458 four hundred fifty-eight

PROBLEM-SOLVING STRATEGY
Make a Table

1. **Read** Ms. Fox saw 5 meteors every minute. How many meteors did she see in 6 minutes?

   Look for a pattern in the table.

   Think Make a table to show that Ms. Fox sees 5 more meteors each minute.

   Write

Minutes	1	2	3	4	5	6
Meteors	5	10	15	20		

   Check Ms. Fox saw ____ meteors in 6 minutes.

2. **Read** There were 12 astronauts. Two went in each . How many had astronauts?

   Think Make a table to show 2 astronauts in each . Stop when you reach 12 astronauts in all.

   Write

Astronauts	2	4	6	8	10	12
	1	2				

   Check There were ____ .
   Each had 2 astronauts.

 Describe the pattern in each table. Name another way to solve each problem.

Read → Think → Write → Check

1. Troy has 70 lights.
   Each model needs 10 lights.
   How many can Troy make?

Lights Needed	10	20	30				
Spaceships Made	1	2					

Troy can make ____ .

2. Lara made 7 constellations.
   Each had 5 ★. Did she draw
   more or less than 32 ★?

Constellations Made	1	2	3				
Stars Used	5	10					

Lara drew ____ than 32 ★.

3. A space probe sends 10 signals every 2 minutes. How many minutes will it take to send 80 signals?

Signals sent	10	20	30	40			
Minutes	2	4					

80 signals will take ____ minutes.

4. Make a table that has a pattern. Then write a problem for the table.

Name _____

PROBLEM-SOLVING APPLICATIONS:
Guess and Test

Read → Think → Write → Check

1. I am a 3-digit number.
   I have less than 4 hundreds.
   I have no ones.
   What number am I?

   I am __210__.

2. I have less than 2 hundreds.
   I have more than 2 ones.
   What number am I?

   I am _____.

3. I am greater than 200.
   I have no tens.
   What number am I?

   I am _____.

4. I have between 2 hundreds
   and 5 hundreds. I have the
   same number of tens as ones.
   What number am I?

   I am _____.

   509 has 5 hundreds 9 ones.

Which numbers in 1–4 have a sum of 509?

509 = _____ + _____

In this lesson your child solved problems by using the Guess and Test strategy.

four hundred sixty-one **461**

Use the steps and strategies you know.

STRATEGY FILE
Choose the Operation
Guess and Check
Make a Table
Extra Information

1. Tara saw 25 ★. Nora saw 9 ★. Nora saw 5 planets. How many fewer stars did Nora see than Tara?

   Nora saw ____ fewer stars.

2. Pierre has 1 ◉. He gets 18 pennies. How much money does he have in all?

   Pierre has _____.

3. At 8:30 Jo saw 5 meteors. Each half hour after, she saw 5 meteors. How many meteors did she see by 10 o'clock?

8:30	9:00		
5			

   She saw ____ meteors.

4. There are 3 constellations. Each has 13 ★. How many ★ is this in all?

   ____ ★ in all

5. There are 34 🛰. 9 of them send signals. How many 🛰 do not send signals?

   ____ 🛰 do not send signals.

6. A 🚀 carried a total of 676 pounds. Color the 2 loads it carried.

   435 pounds   354 pounds
   312 pounds   241 pounds

462 four hundred sixty-two

Chapter Review and Practice

Name _____

Complete the number sentence.

1. $2 + 8 = 40 - \underline{}$
2. $18 - \underline{} = 4 + 5$

Write + or −.

3. $12 \bigcirc 5 \bigcirc 1 = 8$
4. $3 \bigcirc 9 \bigcirc 10 = 22$

Find the missing numbers. ▲ = 5 ◢ = 7

5. ■ − ▲ = 5
   ■ = ___

6. ◢ + ▮ = 15
   ▮ = ___

Add or subtract.

7.
```
  26¢      34      51¢      43      27
+ 15¢    +  8    − 14¢    −  8    −  9
───────  ──────  ───────  ──────  ──────
```

8.
```
 123     303     879     362
+456    +164    −159    −161
─────   ─────   ─────   ─────
```

9.
```
 22      24
 13      22
+41     +30
─────   ─────
```

10. $8 + 2 + 31 = \underline{}$

11. $600 - 400 = \underline{}$

12. I have 3 digits.
    I have more than 2 hundreds.
    I am ___.

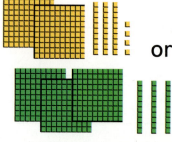

four hundred sixty-three **463**

Name _____

Locate the planets.

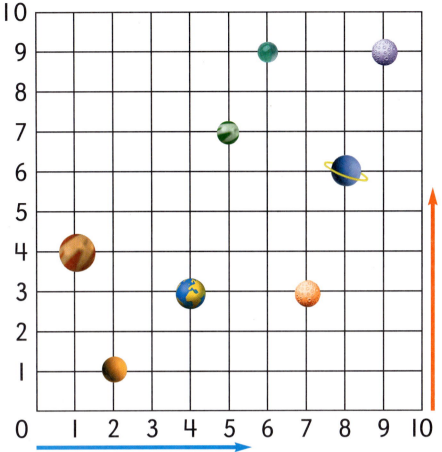

Start at 0.
Count across.
Count up.

🌍 is 4 across and 3 up.

Start at 0.

Complete the chart.

	Across	Up
1. 🪐	7	3
2. 🪐		
3. 🪐		
4. 🪐		
5. 🪐		
6. 🪐		
7. 🪐		

Use the grid above for 8 and 9.

8. Go across 1. Then go up 7. Draw a 🚀.

9. Make a dot on each.

	Across	Up
🔵	4	0
🔴	3	10
🟡	5	5
🟢	9	4
🟣	0	9

Name _____

Performance Assessment

Model with and .

 = 9¢   and   = 3¢

1. 32¢ + = ___

2. 40¢ − = ___

Write + or − to get two different answers.

3. 400 ◯ 200 ◯ 300 = ___
   400 ◯ 200 ◯ 300 = ___

 Choose 1 of these projects.

4. Write two ways to make both sides equal.

  = ___ + ___

 = ___ − ___

5. Copy and complete the tables.

IN	320	705	216
OUT	420	805	

IN	240	175	350
OUT	230	165	

Rule: [IN] + ____

Rule: [IN] − ____

Make a table for [IN] − 100 and for [IN] + 10.

Check Your Mastery

Name _____

Complete the number sentence. | Write + or −.

1. $9 + \underline{} = 15 + 2$
2. $18 - \underline{} = 7 + 3$
3. $4 \bigcirc 8 \bigcirc 2 = 10$
4. $13 \bigcirc 7 \bigcirc 9 = 15$

Find the missing numbers. △ = 8   ▱ = 16

5. ▢ − △ = 9
6. ▱ − 7 = ▢

Find the sum.

7.
```
  42¢        37       703       21       15
+  8¢      +  7     + 125      23       40
                              + 34     + 24
```

8. $7 + 3 + 24 = \underline{}$
9. $1 + 28 + 9 = \underline{}$

Find the difference.

10.
```
  80¢       35¢       44       654      826
−  7¢     − 12¢     −  8     − 324    − 221
```

PROBLEM SOLVING

11. Tara has 1 ▓, 4 ▬, and 2 ▪. Draw the models she needs to show two hundred forty-six.

12. Leon gets 1 (nickel) each day. In how many days can he trade for a quarter?

1	2			

_____ days

466   four hundred sixty-six

Name _____

Still More Practice
CHAPTER 1

Draw. Then write how many.

1. three
2. five
3. two

Draw 1 more ●. Write the number.

4. 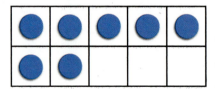 _____    _____

Write the missing number. Write *between*, *after*, *before*.

5. ___, 1   3, ___   9, ___, 11

   ___, 12   8, ___   6, ___, 8

   _____   _____   _____

Model and compare. Write the number.

| six | eight | | eleven | nine |

6. ___ is greater than ___.   7. ___ is less than ___.

   ___ is less than ___.   ___ is greater than ___.

Color fifth red, tenth  blue, and third green.

8.

9. ✔ the odd numbers. [2] [4] [5] [6] [9]

four hundred sixty-seven **467**

Still More Practice
CHAPTER 2

Name _____

Add. Draw to check.

1. 2 + 2 = ___   1 + 1 = ___   3 + 3 = ___

Find the sum. Write the related fact.

2. 1   4   3   5
   +2   +1   +2   +0

3. 2   1   6   3
   +4   +5   +0   +1

Add. Complete each pattern.

4. 0   0
   +6   +4   +___

5. 1   2
   +1   +2   +___

Solve. Write the number sentence.

Favorite Pet

Kind	Number of Pupils
	😀 😀 😀
🐱	😀 😀 😀 😀
🐢	😀
🐦	😀 😀

6. How many liked 🐱 or 🐦?

   ___ ○ ___ = ___

7. How many liked 🐢 or 🐟?

   ___ ○ ___ = ___

468 four hundred sixty-eight

Name _____

Still More Practice
CHAPTER 3

Subtract. Use ■ to model.

1. 6 − 5 = ___   3 − 3 = ___   4 − 2 = ___
2. 5 − 0 = ___   4 − 4 = ___   3 − 1 = ___

Find the difference.

✔ the row that shows a pattern.

3.   2   3   6   4   6   6
   −2   −2   −4   −1   −2   −1

4.   4   5   4   5   6   6
   −3   −5   −0   −1   −6   −0

Write the related number sentence.

5. 1 + 4 = 5
   5 − ___ = ___

6. 3 + 3 = 6
   6 − ___ = ___

7. 2 + 3 = 5
   5 − ___ = ___

8. 3 + 0 = 3
   3 − ___ = ___

9. You have 5 .
You give away 2.
How many are left?

___ ○ ___ = ___

10. Kate draws 2 .
She draws 2 more.
How many in all?

___ ○ ___ = ___

four hundred sixty-nine **469**

Still More Practice
CHAPTER 4

Name _____

Add. Change the order to check mentally.

1. $0 + 8 =$ ___ $9 + 2 =$ ___ $9 + 1 =$ ___
2. $3 + 6 =$ ___ $6 + 4 =$ ___ $7 + 2 =$ ___
3. $2 + 8 =$ ___ $5 + 5 =$ ___ $9 + 3 =$ ___

4. 6   7   5   5   5   6   6
   +2   +3   +3   +6   +2   +6   +1
   ——   ——   ——   ——   ——   ——   ——

5. 8   8   5   4   7   0   4
   +1   +3   +7   +8   +1   +9   +4
   ——   ——   ——   ——   ——   ——   ——

Use a ⟵→ to count on.

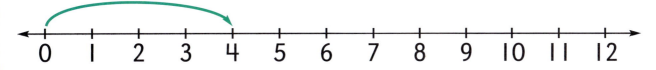

6. $4 + 3 =$ ___   7. $4 + 5 =$ ___   8. $4 + 7 =$ ___

Find the sum. Ring the numbers you add first.

9. $3 + 3 + 5 =$ ___   10. $8 + 2 + 2 =$ ___

   11. Len has 3 🪙. He gets 7¢ more. How much does Len have now?

___¢ ◯ ___¢ = ___¢

470 four hundred seventy

Still More Practice
CHAPTER 5

Name _____

Find the difference.

1. 10 − 9 = ___   9 − 4 = ___   7 − 0 = ___

Subtract. You can check by adding.

2.
```
  9      8     12     10      7      9
 −7     −1     −5     −4     −5     −8
```

3.
```
  8      9     10      8     11      8
 −3     −9     −8     −4     −2     −6
```

4.
```
  7     10     11      9      7     11
 −3     −7     −4     −3     −6     −8
```

Use a ⟵⋯⟶ to subtract.

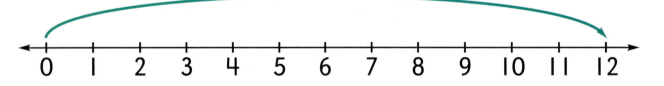

5. 12 − 3 = ___   12 − 6 = ___   12 − 9 = ___

6. Write the family that has more facts.

7. You draw a dozen ☆.
You color 4 blue.
How many are not blue?

___ ◯ ___ = ___

___ ☆ are not blue.

Still More Practice
CHAPTER 6

Name _____

Write how many.

1.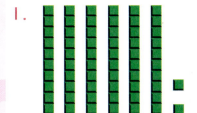

____tens ____ones

____ sixty-_____

2.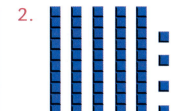

____tens ____ones

____ fifty-_____

3.

____tens ____ones

____ thirty-_____

Write the numbers that each number is between.

4. ___, 40, ___   5. ___, 54, ___   6. ___, 89, ___

Ring the number that is greater. You can model each.

7. | 13 | 31 | 8. | 27 | 17 | 9. | 78 | 87 |

Write the missing numbers.

10. 4, 6, ____, 10, ____, 14, ____, ____, 20

11. 10, 20, ____, 40, ____, ____, 70, ____, ____, 100

12. Make equal groups.

____ groups of 2 = ____

____ twos = ____

13. Show 4 sharing 20 🟩.

____ 🟩 each

14. Which odd numbers between 60 and 80 do you say when you count by 5s? _____

472 four hundred seventy-two

Name _____

Still More Practice
CHAPTER 7

Write how much. ✓ the least amount.

1.  ____ ¢

2. ____ ¢

3.   ____ ¢

Ring the fair trade.

4.     or

Write the time in 2 ways.

5.
half past ____
__:__

6.
____ o'clock
__:__

7.
____ minutes after ____
__:__

March

Sunday	Monday	Tuesday	Wednesday
22	23	24	25
29	30	31	

8. What is the date of the last Monday in March?
_____

9. ✓ more than 1 minute.

read a book

tie your shoes

10. I have a dozen . I lose 5 of them. How much money is left?

___ ◯ ___ = ___

four hundred seventy-three **473**

Still More Practice
CHAPTER 8

Name _____

1. Make a triangle, square, and rectangle.

   ____ sides   ____ corners   ____ sides

   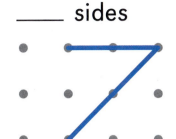   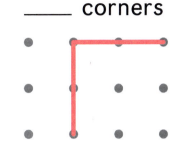   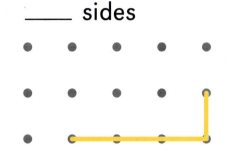

2. Color circles with the same size and shape.

    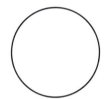

3. Draw to make matching parts.

4. Color pyramid red, rectangular prism blue. ✗ solids that roll.

   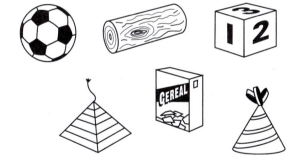

5. Match. Ring the shape with 3 equal parts.

   $\frac{1}{2}$
   $\frac{1}{3}$
   $\frac{1}{4}$

   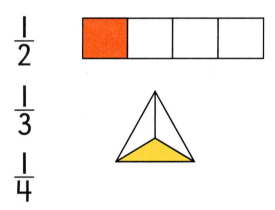

6. Color A so you always get red.

7. Color B so you are less likely to get red.

8. What part of the set in B is red? ____

A   B

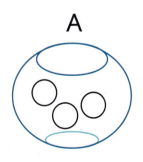   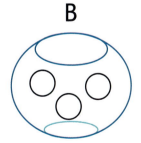

474 four hundred seventy-four

Name _____

Still More Practice
CHAPTER 9

Find the sum or difference.

1.   20   51   44   10   73   25
   +40   +23   + 5   +22   +14   +20

2.   50   74   66   37   49   28
   −10   −13   −20   − 2   −15   −18

Add or subtract. Watch for + and −.

3.   25¢   92¢   35¢   78¢   17¢
   +41¢   −22¢   +24¢   −40¢   +62¢

Estimate the answer.

4. 37 + 24 is about ____.   5. 58 − 29 is about ____.

6. 22 + 47 is about ____.   7. 83 − 14 is about ____.

Add or subtract. Regroup if needed.

8. 3 tens 6 ones + 5 ones   9. 6 tens 2 ones − 4 ones

   ____tens ____one   ____tens ____ones

PROBLEM SOLVING

10. Ann has 27¢. She finds 12¢. Then Ann loses 10¢. How much does Ann have now?

    _____

11. One of my 5 coins is a nickel. I have the same number of dimes as pennies. How much do I have?

    _____

four hundred seventy-five **475**

Still More Practice
CHAPTER 10

Name _____

Use your ruler. Measure.

1.

2.

____ inch   ____ centimeters

Which holds more? Ring.

3.

4.  or

5. Write more or less.

I foot	I pound	I liter	I kilogram

_____   _____   _____   _____

6. Which would you use to measure?

A    B   C

        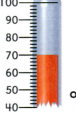   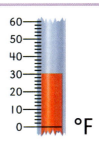

____   ____   ____

7. **PROBLEM SOLVING** Which shows the temperature for sledding? Ring.

476 four hundred seventy-six

Still More Practice
CHAPTER 11

Name _____

Find the sum or difference.

1.   5   7   9   4   7   8   9
   +8   +7   +8   +9   +9   +7   +9

2.   8¢   7¢   9¢   5¢   8¢   8¢
   +6¢   +8¢   +7¢   +9¢   +9¢   +5¢

3. 14   13   16   15   16   15   17
   − 5   − 8   − 7   − 6   − 8   − 9   − 9

4. 16¢   15¢   14¢   14¢   13¢   14¢
   − 9¢   − 7¢   − 8¢   − 6¢   − 4¢   − 9¢

Write the fact family.

5.  ___ + ___ = ___   ___ + ___ = ___

     ___ − ___ = ___   ___ − ___ = ___

Add.

6.   8   4   3   9   4   6   3
     2   5   4   2   4   3   3
   +8   +5   +3   +2   +8   +6   +8

7. I start with 18 🐚. I keep as many as I give away. How many do I have now? _____

four hundred seventy-seven

MENTAL MATH

Set 1
1 more | **1 fewer**

Set 2
Count on | **Count back**

1 to 5 | 5 to 1
3 to 7 | 7 to 3
0 to 4 | 4 to 0
4 to 8 | 8 to 4
6 to 12 | 12 to 6
2 to 6 | 6 to 2

Set 3
Between

one ___ three
zero ___ two
four ___ six
five ___ seven
ten ___ twelve
seven ___ nine

Set 4
Before | **After**

___, 4 | 10, ___
___, 3 | 7, ___
___, 12 | 0, ___
___, 2 | 11, ___
___, 5 | 9, ___
___, 10 | 6, ___

Set 5

Set 6

1 + 1 | 3 + 2
3 + 1 | 0 + 5
2 + 2 | 5 + 1
3 + 0 | 2 + 1
4 + 1 | 3 + 3
1 + 2 | 4 + 2

Set 7
Add

0 to: 1, 2, 3, 4, 5, 6
1 to: 0, 1, 2, 3, 4, 5
2 to: 0, 1, 2, 3, 4
3 to: 0, 1, 2, 3
4 to: 0, 1, 2
5 to: 0, 1

Set 8
How many in all?

2 and 2
3 and 2
1 and 4
4 and 2
3 and 3
5 and 1

Set 9

3 − 1 | 5 − 2
4 − 2 | 6 − 6
2 − 1 | 5 − 3
3 − 2 | 6 − 2
4 − 1 | 5 − 1
1 − 0 | 6 − 3

Set 10
Subtract ... from

0	6, 5, 4, 3, 2, 1
1	6, 5, 4, 3, 2, 1
2	6, 5, 4, 3, 2
3	6, 5, 4, 3
4	6, 5, 4
5	6, 5

Set 11
How many left?

Have	Lose
5	4
3	1
4	2
2	2
5	3
6	3

Set 12

$4 + 2$	$6 + 3$
$6 + 1$	$5 + 5$
$4 + 4$	$7 + 2$
$6 + 2$	$5 + 4$
$4 + 3$	$6 + 4$
$5 + 3$	$7 + 3$

Set 13

$8 \\ +2$	$7 \\ +3$	$8 \\ +4$
$9 \\ +3$	$6 \\ +3$	$7 \\ +4$
$8 \\ +3$	$6 \\ +6$	$5 \\ +7$

Set 14
Add ... to

1	1, 3, 5, 7, 9, 8, 6
2	2, 4, 6, 8, 10, 9
3	0, 2, 4, 6, 8, 9, 7
4	1, 3, 5, 7, 8, 6, 4
5	10, 2, 3, 4, 5, 6, 7
6	6, 5, 4, 3, 2, 1, 0

Set 15
How many in all?

5, 5 more
7, 3 more
9, 3 more
6, 4 more
8, 2 more
9, 2 more

Set 16

8 2 +1 2 8 +2	7 3 +1 3 7 +2	6 4 +1 4 6 +2

Double. Then add 2.
6, 5, 4, 3, 2, 1

Set 17

$7 - 4$	$9 - 2$
$8 - 3$	$10 - 9$
$7 - 6$	$9 - 6$
$7 - 2$	$9 - 4$
$8 - 5$	$10 - 8$
$8 - 4$	$10 - 6$

Set 18

$12 \\ -3$	$11 \\ -4$	$11 \\ -6$
$12 \\ -5$	$12 \\ -7$	$11 \\ -5$
$12 \\ -4$	$11 \\ -3$	$12 \\ -6$

MAINTENANCE

Mental Math

Set 19

(dominoes)

Set 20

9	9+0, 8+1, 7+2,...
10	10+0, 9+1, 8+2,...
11	11+0, 10+1, 9+2,...
12	12+0, 11+1, 10+2,...

4	4−0, 5−1, 6−2,...
5	5−0, 6−1, 7−2,...
6	6−0, 7−1, 8−2,...

Set 21

1 ten 6 ones
2 tens 7 ones
5 tens 4 ones
8 tens 3 ones

17	27	53	45
18	16	92	69
40	20	38	71
35	25	93	99

Set 22

Count by	from	to
1	62	82
2	24	44
5	30	60
10	30	100
5	55	95
2	76	96

Set 23

4 (nickels) 3 (nickels) 2 (pennies)
7 (nickels) 2 (nickels) 3 (pennies)
3 (nickel) 1 (dime) 8 (pennies)
7 (dime) 5 (dime) 4 (pennies)
5 (dime) 1 (quarter) 1 (penny)

Set 24

2 (quarter) 1 (nickel) 1 (penny)
3 (quarter) 2 (nickel) 2 (penny)
4 (quarter) 3 (nickel) 1 (penny)
1 (quarter) 1 (nickel) 2 (penny)
1 (quarter) 1 (nickel) 1 (penny)
1 (quarter) 1 (nickel) 1 (penny)

Set 25

(clocks)

Set 26

12:00 1:00   2:00
8:00   9:00   10:00
10:30 11:30 12:30
3:30   4:30   5:30
7:30   8:00   8:30
11:00 11:30 12:00

Set 27

(shapes: circle, cube, square, cylinder, rectangle, sphere, triangle, cone, hexagon, rectangular prism)

Set 28

Set 29

Set 30

50	40	70
+20	+40	+10

80	70	90
−30	−40	−60

30 + 20	40 − 10
50 + 40	90 − 30
20 + 20	70 − 20

Set 31

10 + 6	36 − 5
20 + 5	68 − 4
40 + 2	59 − 6
71 + 10	52 − 20
63 + 20	32 − 10
59 + 30	55 − 40

Set 32

Add ... to

2	21, 22, 23, 24
4	55, 44, 33, 22
6	63, 62, 61, 60
5	24, 33, 42, 51
3	16, 15, 14, 13
1	98, 87, 76, 65

Set 33

Subtract ... from

2	29, 38, 47, 56
3	99, 88, 77, 66
4	54, 55, 56, 57
5	38, 48, 58, 68
6	76, 77, 78, 79
7	99, 87, 79, 67

Set 34

Double ... then

	add	subtract
20¢	2¢	10¢
40¢	5¢	20¢
30¢	4¢	10¢
22¢	2¢	3¢
11¢	6¢	2¢
13¢	5¢	4¢

Set 35

Regroup: ? tens ? ones

16 ones	11 ones
23 ones	19 ones
17 ones	20 ones

Regroup: ? ones

1 ten 2 ones
1 ten 5 ones
1 ten 8 ones
1 ten 4 ones

Set 36

inches   kilograms
pint   centimeters
feet   pounds

four hundred eighty-one 481

Mental Math

Set 37

7 +6	7 +7	8 +6
9 +4	5 +8	9 +5
9 +8	8 +8	7 +9

Set 38

14 − 3	13 − 7	14 − 6
13 − 5	14 − 7	15 − 9
16 − 7	18 − 9	15 − 8

Set 39

5 4 +5	9 3 +6	8 5 +2

9 + 1 + 7
6 + 6 + 4
8 + 8 + 1
3 + 5 + 7
5 + 9 + 5

Set 40

I had	I got
4¢ + 4¢	4¢ more
7¢ + 7¢	2¢ more
6¢ + 6¢	1¢ more
3¢ + 3¢	5¢ more
5¢ + 5¢	7¢ more
8¢ + 8¢	1¢ more

Set 41

I had	I spent
6¢ + 4¢	3¢
7¢ + 9¢	8¢
6¢ + 8¢	7¢
9¢ + 6¢	8¢
7¢ + 5¢	9¢
8¢ + 5¢	7¢

Set 42

5, 15, 25, 35, 45,…	+5
6, 16, 26, 36, 46,…	+6
7, 17, 27, 37, 47,…	+7
8, 18, 28, 38, 48,…	+8
12, 22, 32, 42, 52,…	−3
11, 21, 31, 41, 51,…	−4
10, 20, 30, 40, 50,…	−5

Set 43

+ or −	equals
8 ○ 2 ○ 1	11, 5
10 ○ 5 ○ 2	7, 3
4 ○ 4 ○ 2	10, 6
9 ○ 1 ○ 3	13, 5
12 ○ 6 ○ 1	5, 7
10 ○ 1 ○ 2	7, 11

Set 44

△ = 1   □ = 10

14 − □	28 − □
25 + △	20 + △
16 + □	45 − □
14 − △	19 + △

20 + □ − △
50 − □ + △

Set 45

32 + □ = 32
□ + 10 = 46
17 + □ = 18
41 − □ = 41
15 − □ = 0
37 − □ = 27
20 − □ = 19

GLOSSARY

This magazine belongs to

_____.

addition joining

___ + ___ = ___

___ and ___ equals ___ in all.

addends ④ + ②

after 16, ⑰

balance

bar graph

Shapes

before ⑫, 13

between 21, ㉒, 23

---

side

skip count by 5 0, 5, 10, 15, 20, 25,…
by 10 0, 10, 20, 30, 40, 50,…

square

sphere

subtraction  separating
Take away 2.
3 left.

sum 3 + 4 = ⑦

symmetry

tally marks ||||̸ | six

triangle

GLOSSARY

calculator

calendar → month

cent sign ¢

centimeter

circle

clock — minute hand, hour hand

cone

corner

cube

---
fold

---

pint

place value — 2 tens 3 ones

plus + plus sign

pound

pyramid

quart + =

quarter 25¢ or

rectangle

rectangular prism

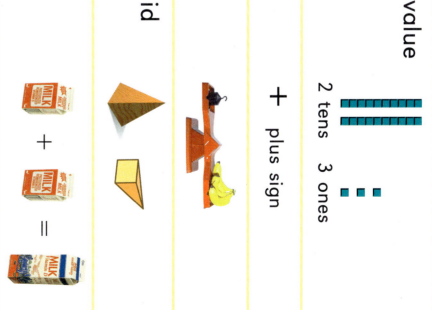